# Nuclear Magnetic Resonance Spectroscopy

# Analytical Chemistry by Open Learning

**Titles in Series:**

Samples and Standards
Sample Pretreatment
Classical Methods
Measurement, Statistics and Computation
Using Literature
Instrumentation
Chromatographic Separations
Gas Chromatography
High Performance Liquid Chromatography
Electrophoresis
Thin Layer Chromatography
Visible and Ultraviolet Spectroscopy
Fluorescence and Phosphorescence Spectroscopy
Infra Red Spectroscopy
Atomic Absorption and Emission Spectroscopy
Nuclear Magnetic Resonance Spectroscopy
X-Ray Methods
Mass Spectrometry
Scanning Electron Microscopy and Microanalysis
Principles of Electroanalytical Methods
Potentiometry and Ion Selective Electrodes
Polarography and Other Voltammetric Methods
Radiochemical Methods
Clinical Specimens
Diagnostic Enzymology
Quantitative Bioassay
Assessment and Control of Biochemical Methods
Thermal Methods
Microprocessor Applications

# Nuclear Magnetic Resonance Spectroscopy

Analytical Chemistry by Open Learning

Author:
DAVID A R WILLIAMS
*Manchester Polytechnic*

Editor:
DAVID J MOWTHORPE
*Sheffield City Polytechnic*

*on behalf of ACOL*

Published on behalf of ACOL, London
by
JOHN WILEY & SONS
Chichester · New York · Brisbane · Toronto · Singapore

Published by permission of the Controller of
Her Majesty's Stationery Office



***Library of Congress Cataloging in Publication Data:***

Williams, David A. R.
Nuclear magnetic resonance spectroscopy.

Includes index.
1. Nuclear magnetic resonance spectroscopy.
I. Mowthorpe, David J. II. ACOL (Firm: London,
England) III. Title.
QD96.N8W55 1986 543'.0877 86-15694
ISBN 0 471 91176 3
ISBN 0 471 91177 1 (pbk.)

***British Library Cataloguing in Publication Data:***

Williams, David A. R.
Nuclear magnetic resonance spectroscopy.—(Analytical chemistry)
1. Nuclear magnetic resonance spectroscopy
I. Title II. Series
543'.0877 QD96.N8

ISBN 0 471 91176 3 (Cloth)
ISBN 0 471 91177 1 (Paper)

Printed and bound in Great Britain

# Analytical Chemistry

This series of texts is a result of an initiative by the Committee of Heads of Polytechnic Chemistry Departments in the United Kingdom. A project team based at Thames Polytechnic using funds available from the Manpower Services Commission 'Open Tech' Project have organised and managed the development of the material suitable for use by 'Distance Learners'. The contents of the various units have been identified, planned and written almost exclusively by groups of polytechnic staff, who are both expert in the subject area and are currently teaching in analytical chemistry.

The texts are for those interested in the basics of analytical chemistry and instrumental techniques who wish to study in a more flexible way than traditional institute attendance or to augment such attendance. A series of these units may be used by those undertaking courses leading to BTEC (levels IV and V), Royal Society of Chemistry (Certificates of Applied Chemistry) or other qualifications. The level is thus that of Senior Technician.

It is emphasised however that whilst the theoretical aspects of analytical chemistry can be studied in this way there is no substitute for the laboratory to learn the associated practical skills. In the U.K. there are nominated Polytechnics, Colleges and other Institutions who offer tutorial and practical support to achieve the practical objectives identified within each text. It is expected that many institutions worldwide will also provide such support.

The project will continue at Thames Polytechnic to support these 'Open Learning Texts', to continually refresh and update the material and to extend its coverage.

Further information about nominated support centres, the material or open learning techniques may be obtained from the project office at Thames Polytechnic, ACOL, Wellington St., Woolwich, London, SE18 6PF.

# How to Use an Open Learning Text

Open learning texts are designed as a convenient and flexible way of studying for people who, for a variety of reasons cannot use conventional education courses. You will learn from this text the principles of one subject in Analytical Chemistry, but only by putting this knowledge into practice, under professional supervision, will you gain a full understanding of the analytical techniques described.

To achieve the full benefit from an open learning text you need to plan your place and time of study.

- Find the most suitable place to study where you can work without disturbance.

- If you have a tutor supervising your study discuss with him, or her, the date by which you should have completed this text.

- Some people study perfectly well in irregular bursts, however most students find that setting aside a certain number of hours each day is the most satisfactory method. It is for you to decide which pattern of study suits you best.

- If you decide to study for several hours at once, take short breaks of five or ten minutes every half hour or so. You will find that this method maintains a higher overall level of concentration.

Before you begin a detailed reading of the text, familiarise yourself with the general layout of the material. Have a look at the course contents list at the front of the book and flip through the pages to get a general impression of the way the subject is dealt with. You will find that there is space on the pages to make comments alongside the text as you study—your own notes for highlighting points that you feel are particularly important. Indicate in the margin the points you would like to discuss further with a tutor or fellow student. When you come to revise, these personal study notes will be very useful.

Π When you find a paragraph in the text marked with a symbol such as is shown here, this is where you get involved. At this point you are directed to do things: draw graphs, answer questions, perform calculations, etc. Do make an attempt at these activities. If necessary cover the succeeding response with a piece of paper until you are ready to read on. This is an opportunity for you to learn by participating in the subject and although the text continues by discussing your response, there is no better way to learn than by working things out for yourself.

We have introduced self assessment questions (SAQ) at appropriate places in the text. These SAQs provide for you a way of finding out if you understand what you have just been studying. There is space on the page for your answer and for any comments you want to add after reading the author's response. You will find the author's response to each SAQ at the end of the text. Compare what you have written with the response provided and read the discussion and advice.

At intervals in the text you will find a Summary and List of Objectives. The Summary will emphasise the important points covered by the material you have just read and the Objectives will give you a checklist of tasks you should then be able to achieve.

You can revise the Unit, perhaps for a formal examination, by re-reading the Summary and the Objectives, and by working through some of the SAQs. This should quickly alert you to areas of the text that need further study.

At the end of the book you will find four reference lists of commonly used scientific symbols and values, units of measurement and also a periodic table.

# Contents

# Study Guide

This short course is designed to introduce you to some of the most important aspects of nuclear magnetic resonance that are in widespread use in both industrial and academic laboratories. Proton and carbon-13 are by far the most widely used nuclei in nmr spectroscopy, and as such are dealt with almost exclusively. The principles of nmr are common to all nuclei and hence if, for example, you are interested in using fluorine-19 or phosphorus-31 nmr, little further study would be needed.

From our experience of teaching nmr spectroscopy over a number of years, we have found that the major difficulty students experience is the large number of new ideas. Hence a course such as this, which allows you to study at your own pace, is particularly beneficial.

The course is biased towards applications at the expense of some theory. In this respect it differs from many other publications on the subject. You will find, however, it provides an adequate theoretical basis for dealing with spectral interpretation and quantitative analysis. If you get a chance to carry out some laboratory work in conjunction with this course, do so. It will help a great deal in developing your understanding and expertise.

Although you do not need any prior knowledge of nmr before embarking on this course, it is advantageous for you to have studied chemistry since leaving school. Typically, you will have studied chemistry for a further two or three years together with supporting studies in physics and mathematics. More specifically you will need to be familiar with structural organic chemistry, the electronmagnetic spectrum, any branch of spectroscopy, and basic analytical chemistry.

# Practical Objectives

A short laboratory course in nmr spectroscopy would benefit any student studying this Unit.

The successful attainment of the three objectives given below would take a minimum of six hours of practical work. Such work would directly support Parts 2, 4 and 5 of the text. Practical work is not of direct relevance to the other Parts, 1 and 3.

By the end of a practical course the student should be able to:

1. Prepare a solid or liquid sample for nmr analysis by selection of a suitable solvent, etc.

2. Obtain a routine CW nmr spectrum and its integration curve from an analytical sample, by selecting appropriate instrumental conditions.

3. Analyse quantitively, via integration data, samples such as those examined in Part 5 of the text.

# Bibliography


1. (*a*) F W Fifield and D Kealey, *Principles and Practice of Analytical Chemistry*, 2nd ed., International Text-book Co.

   (*b*) D A Skoog and D M West, *Principles of Instrumental Analysis*, 4th ed., Holt, Rinehart and Winston, 198

   (*c*) H H Willard, L L Merritt, J A Dean, F A Settle, *Instrumental Methods of Analysis*, 6th ed., Wadsworth Publishing Co., 1981.

2. F Kasler, *Quantitative Analysis by NMR Spectroscopy*, Academic Press, 1973.

3. R J Fessenden, J S Fessenden, *Organic Chemistry*, 2nd ed., Willard Grant Press, 1982. Or any other general organic text.

4. R David, D H J Wells, *Spectral Problems in Organic Chemistry*, International Textbook Co., 1984.

5. D E Leyden, R H Cox, *Analytical Applications of NMR*, Wiley Interscience, 1977.

6. M L Martin, J J Delpuech, G J Martin, *Practical NMR Spectroscopy* Heyden, 1980.

7. F W Wehrli, T Wirthlin, *Interpretation of Carbon-13 NMR Spectra*, John Wiley and Sons, 1983.

8. H Günther, *NMR Spectroscopy*, Wiley, 1980.

9. R K Harris, *Nuclear Magnetic Resonance Spectroscopy; A Physico-Chemical Approach*, Pitman, 1983.

10. J W Akitt, *NMR and Chemistry*, 2nd ed., Chapman Hall, 1983

11. E Breitmaier, G Haas, W Voelter, *Atlas of Carbon-13 NMR Data*, Heyden, 1979.

**NOTE** References 1–4 are books immediately useful at this level: Reference 1 gives three books which contain a single chapter devoted to nmr. References 5–8 are more advanced, but some chapters are directly relevant; Reference 9 and 10 are of Honours level and beyond; Reference 11 is the source of the carbon-13 data used in this Unit.

# Acknowledgements

Fig. 3.1d redrawn with permission from F. W. Wehrli and T. Wirthlin, *Interpretation of Carbon-13 NMR Spectra*, John Wiley & Sons Ltd, Chichester (1976).

Fig. 4.3c redrawn with permission from J. K. Akett, *NMR and Chemistry*, 2nd Edn, Chapman & Hall (1983).

The author would also like to thank Paul Warren for his many helpful discussions.

# 1. Background and Theory of NMR Spectroscopy

## Overview

In this first part of the course we are going to study quite a number of ideas about nuclear magnetic resonance spectroscopy. As you will appreciate later in this course and if you go on to study this fascinating and wide-ranging technique at higher levels you need to have a sound knowledge of the physical origin of the nmr phenomenon and its associated parameters. So we start off by looking at spin-active nuclei and how nuclei can absorb radiofrequency radiation—the physical basis of nmr.

We shall encounter a couple of fundamental equations which govern some nmr processes and we will explore the information that nmr spectroscopy can give us in terms of four parameters measurable from an nmr spectrum: chemical shifts, coupling constants, relaxation times, and integrations. You will be able to explain and exemplify all these terms by the end of this part of the Unit.

We shall see that all of these parameters are closely linked with molecular structure. Chemical shifts of protons and carbon-13 nuclei in organic structures will be explained in terms of bonding and electronic effects and you will encounter the mysteries of magnetic anisotropy (pronounced an-isotropy).

The fine structure associated with many nmr spectra will be seen to be related to spin-spin coupling and can give information about neighbouring nuclei. An important rule—the $2nI + 1$ rule—will be introduced and developed.

When it comes to relaxation times you won't be doing any relaxation—only the spin-active nuclei. This parameter affects the width of nmr signals and has other far reaching consequences.

Finally, in this part of the Unit you will learn about integration of nmr signals and how the area under a signal is proportional to the number of spin-active nuclei causing the signal. This lays the basis for quantitative analysis (Part 5) as well as being an aid in qualitative work (Part 3).

Don't worry if you find this first part tough going because, like all of the Units of the Analytical Chemistry series, you can study at your own pace and come back to parts you do not follow, later on.

## 1.1 NUCLEAR MAGNETIC RESONANCE AND SPIN-ACTIVE NUCLEI

When certain types of nuclei like protons, carbon-13 and fluorine-19 are placed in a strong magnetic field they can absorb electromagnetic radiation in the radiofrequency range. Such nuclei are said to be *spin-active* and to *resonate*. The precise frequencies at which spin-active nuclei resonate can be picked up and displayed by instruments called *nuclear magnetic resonance* (nmr) spectrometers.

To explain the phenomenon of nmr we must look at the nature of a spin-active nucleus such as a proton. As well as having an electrical charge associated with it, a nucleus behaves as if it were *spinning* about an axis (a bit like a gyroscope in motion or the Earth spinning on its axis). Fig. 1.1a shows the idea. Now, charged spinning particles generate a magnetic field around them, so that we can think of protons and so on as tiny spinning magnets (microscopic compass needles). In the absence of any external magnetic field these nuclei are spinning at random in their atomic or molecular environment, but when placed in a strong external field ($B_O$) these nuclear magnets orientate themselves with respect to the direction of the magnetic field, again just like compass needles align themselves in the Earth's magnetic field. Unlike compass needles, though, spin-active nuclei can orientate themselves in more than one way. For protons there are two possible alignments, either with the field, or against it (Fig. 1.1a).

The two orientations of our example are also known as the $\alpha$ and $\beta$ *spin states*, and these states differ very slightly in energy. It is this energy difference that can be supplied by the radiofrequency radiation allowing the nuclear spins to change their state. (Fig. 1.1b).

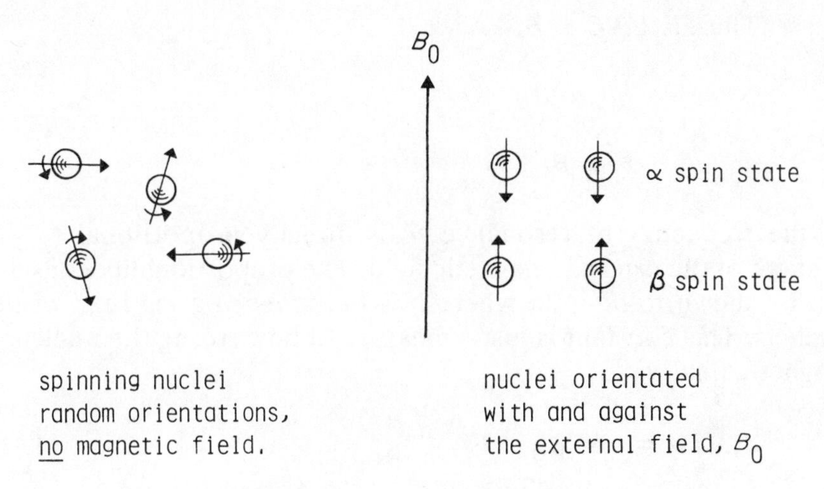


**Fig. 1.1a.** *Nuclear spin*

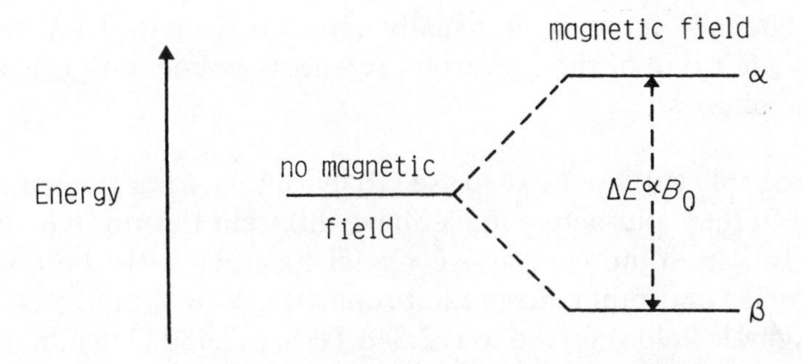


**Fig. 1.1b.** *Energy description of the nmr phenomenon*

A fundamental observation about the nmr phenomenon is that the energy difference ($\Delta E$) between spin states is directly proportional to the magnetic field strength. From this observation we can derive a very important, yet fairly simple equation governing nmr. According to Planck's Law:

$$\Delta E = h\nu \tag{1.1a}$$

where $h$ is Planck's constant, and

$\nu$ is a frequency of electromagnetic radiation.

Thus as $\Delta E \propto B_O$

$h\nu \propto B_O$

or $\nu \propto B_O$

ie, the frequency of resonance $\nu$, is directly proportional to the strength of the external magnetic field. The proportionality constant can be shown to be $\gamma/2\pi$ where $\gamma$ is the *magnetogyric ratio* of the nucleus. This constant is just a measure of how strong the nucleus's magnetic field is.

$$\text{Thus } \nu = \frac{\gamma}{2\pi} B_O \tag{1.1b}$$

We will come across this equation in one form or another throughout this Unit.

The frequency term, $\nu$, is usually referred to as the Larmor frequency after one of the important scientists linked with this area of nuclear physics.

You probably will not be surprised to learn that, for a given magnetic field, different spin-active nuclei have different Larmor frequencies. Fig. 1.1c lists some spin-active nuclei together with their Larmor frequencies and other magnetic properties. You should notice that the magnetic field referred to is 2.348 Tesla (23,480 Gauss in non-S.I. units).

This is a considerable strength for a magnet when you realise that the Earth's magnetic field is only about 1–2 × $10^{-4}$ Tesla (1–2 Gauss). We will be talking more about magnets in Part 2.

To highlight some of the points noted above I shall pick two examples from Fig. 1.1c. The Larmor frequency for the proton is 100 MHz (100 million Hertz), a little bit more than VHF radio, while at the same magnetic field strength of 2.348 Tesla carbon-13 has a Larmor frequency of 25.1 MHz.

| Nucleus | $^{1}$H | $^{2}$H | $^{13}$C | $^{19}$F | $^{31}$P |
|---|---|---|---|---|---|
| Larmor Frequency*/MHz | 100 | 15.3 | 25.1 | 94.1 | 40.5 |
| Sensitivity** | 1 | $1.5 \times 10^{-6}$ | $1.8 \times 10^{-4}$ | 0.83 | 0.07 |
| Spin Quantum Number ($I$) | $\frac{1}{2}$ | 1 | $\frac{1}{2}$ | $\frac{1}{2}$ | $\frac{1}{2}$ |
| % Natural Abundance | 99.985 | 0.015 | 1.1 | 100 | 100 |

* For magnetic field strength of 2.348 Tesla

** This figure is proportional to the cube of the magnetogyric ratio and is a measure of the relative sensitivity of the nucleus at natural abundance.

**Fig. 1.1c.** *Magnetic properties of some spin-active nuclei*

You may be wondering why we must have such powerful magnetic fields in order to observe nmr spectra. Why can we not just use the Earth's magnetic field? The answer is to be found in the Boltzmann Distribution, a mathematical expression which links the populations of different energy levels to the energy difference between them. One form of this mathematical equation is:

$$N_2/N_1 = 1 - \Delta E/kT \qquad (1.1c)$$

where $N_2$ and $N_1$ are the populations of the upper and lower energy levels (spin states for nmr); $\Delta E$ is the energy difference between the energy levels; $k$ is a constant, the Boltzmann constant, and $T$ is the absolute temperature.

Now the energy difference for spin states $\alpha$ and $\beta$ is small so the term $\Delta E/kT$ is also small. Thus $N_2/N_1$ is almost unity, ie there is only a tiny population difference between the two spin states in favour of $N_1$. If you look at Fig. 1.1b you can see what can be done to increase $\Delta E$, and so increase the population difference. That's right, we need to increase $B_0$, the magnetic field as $\Delta E \propto B_0$. As we shall see later we need as big a population difference as possible to obtain suitably powerful nmr signals and so we need intense, powerful magnetic fields. As with Eq. 1.1b we are going to find the Boltzmann equation popping up in several places in this Unit.

Fig. 1.1c also prompts the question of which elements have spin-active nuclei. In fact only nuclei that have a spin quantum number ($I$) not equal to zero can be spin-active. So carbon-13 has a spin quantum number of $\frac{1}{2}$ and is spin-active while carbon-12 the most abundant carbon isotope (98.8%) has a spin quantum number of zero and is not spin-active. Perhaps that is just as well because if carbon-12 were to be active, nmr spectra of organic compounds would turn out to be hopelessly complex.

Most elements have some spin-active isotopes and you could consult reference 9 or 10 of the Bibliography to find full 'spin-active' periodic tables.

Before we progress to the next section, a final point about spin quantum numbers; we can think of spin quantum numbers as a measure of how many different ways the nuclear magnets can align themselves in a magnetic field. The actual relationship is given by the expression $(2I + 1)$ where $I$, the spin quantum number, can have half-integer or integer values ($\frac{1}{2}$, 1, $\frac{3}{2}$, etc). So for the proton with $I = \frac{1}{2}$ there must be $(2 \times \frac{1}{2} + 1)$ spin states, ie, 2. For deuterium, hydrogen-2, $I = 1$ so there must be 3 spin states or alignments in a magnetic field. The consequences of this idea are explored in greater detail in the section on spin–spin coupling.

**SAQ 1.1a**

Which of the following statements about the nmr phenomenon are correct?

Circle T (true) or F (false).

(*i*) Nuclear magnetic resonance is to do with the absorption of radiowaves by certain kinds of nuclei when they are in a strong magnetic field.

T / F

(*ii*) The precise frequency that a nucleus will absorb, when placed in a magnetic field is given by the equation $\nu = \frac{\gamma}{2\pi} B_0$

T / F

(*iii*) All nuclei can undergo nmr.

T / F

(*iv*) The number of spin states for a nucleus is given by the expression $(2I + 1)$ where $I$ is the spin quantum number of the nucleus.

T / F

**SAQ 1.1b**

Boron-11 has a spin quantum number of $\frac{3}{2}$. How many spin states can it have?

(*i*) 6

(*ii*) 5

(*iii*) 4

(*iv*) 3

**SAQ 1.1c**

From the list of words/symbols fill in the spaces (1) to (4) in the following diagram which is an energy description of the nmr phenomenon. Note that there are some items in the list that *do not* fit at all.

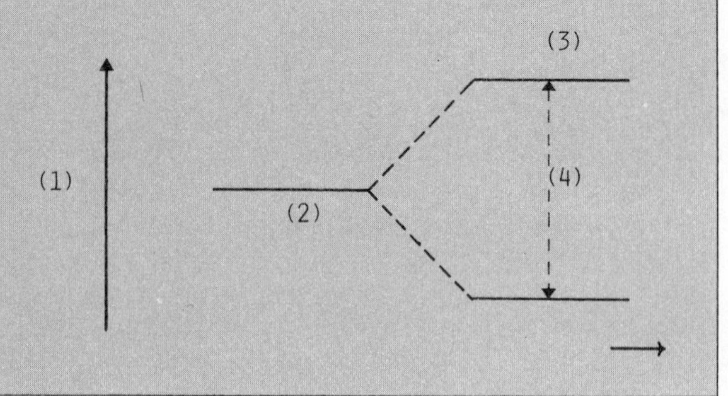

**SAQ 1.1c (cont.)**

(*i*) $\Delta E \propto \Delta B_O$

(*ii*) Energy

(*iii*) magnetic field

(*iv*) $h\nu \propto B_O$

(*v*) zero field

(*vi*) $\nu \propto \gamma/2\pi$

**SAQ 1.1d**

Given the Boltzmann constant to be

$1.3806 \times 10^{-23}$ J K$^{-1}$

and Planck's constant to be

$6.6262 \times 10^{-34}$ Js

which of the following is the population difference between the lower and upper spin states for spin-active nuclei resonating at 100 MHz and 33 °C in a 2.348 Tesla field?

(*i*) about 7 per million

(*ii*) about 9 per million

(*iii*) about 16 per million

(*iv*) about 145 per million

## 1.2. NMR PARAMETERS

In the previous section you saw how a spin-active nucleus like the proton when placed in a magnetic field, $B_O$, could resonate upon irradiation with radiofrequency waves at its Larmor frequency. In this section we will look at the *four* nmr parameters that can be measured from nmr spectra and which can be used in quantative and/or qualitative analysis. These four parameters are ***chemical shifts, spin–spin coupling constants, relaxation times, and integrations.***

Fig. 1.2a summarises the units, symbols and analytical potential for these parameters. The following sections explain these parameters in greater detail.

| Parameter | Units | Main Analytical Use* | Symbol(s) |
|---|---|---|---|
| Chemical Shift | ppm | QL | $\delta$ |
| Coupling Constant | Hz | QL | J |
| Relaxation Time | s | QL/QN | $T_1$, $T_2$ |
| Integration | – | QN | – |

* QN = quantitative
QL = qualitative

**Fig. 1.2a**

## 1.3. CHEMICAL SHIFTS

For an isolated nucleus, for example carbon-13, resonance will occur at its Larmor frequency. However, in a molecule the carbon atom can be bonded to other carbons, hydrogen, and many other elements. The electrons in these bonds and elements influence the local magnetic field around the nucleus in question so modifying in size the external magnetic field, $B_O$, felt by the nucleus. Thus the Larmor frequencies for such nuclei will be different from that of an isolated nucleus.

Electrons, whether from bonding or non-bonding orbitals, *shield* the nucleus from $B_O$. They do so because electrons in a molecule are responsible for small magnetic fields which oppose the external magnetic field, $B_O$, at a nucleus. The size of this small shielding magnetic field will vary with the electron density at a particular nucleus, which is of course dependent on molecular structure. Hence chemically different nuclei will be shielded to different extents and therefore resonate at slightly different frequencies.

In practice then we will see a set of resonance absorptions in an nmr spectrum corresponding to all the chemically different spin-active nuclei. It is very difficult to measure *absolute* frequencies in nmr spectroscopy, but it is much easier, experimentally, to measure very precise frequency differences between resonances. Consequently, the positions of these resonances in the spectrum are usually measured with respect to their *shift* from a standard which has been added to the sample. The usual standard for proton and carbon-13 spectra is tetramethylsilane (TMS) whose proton and carbon resonances are highly shielded and are usually defined as the zero point of the spectrum (Fig. 1.3a). Thus the set of resonances are said to have *chemical shifts* with respect to TMS. (*Chemically* different nuclei whose resonances are *shifted* from that of the standard).

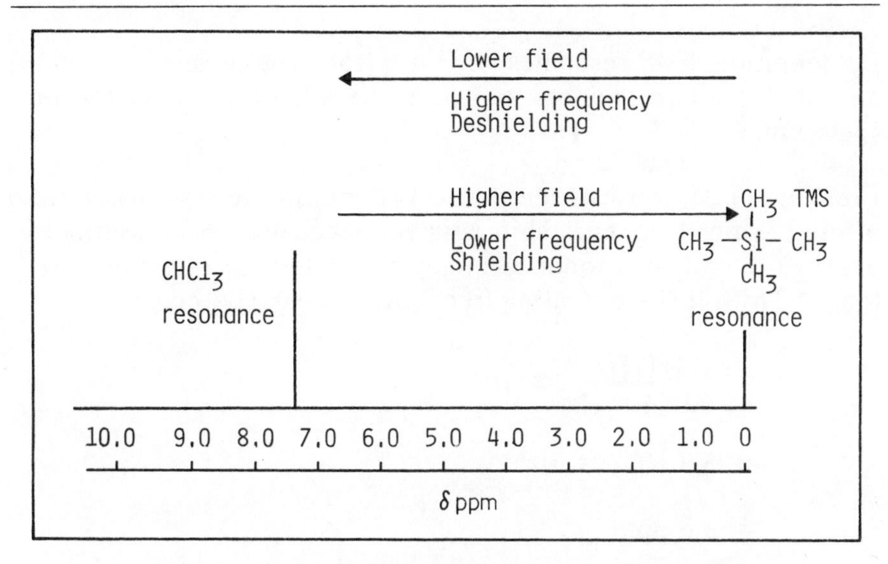


**Fig. 1.3a.** ***Relationship between chemical shifts, frequency, magnetic field and shielding in an nmr spectrum***

TMS is a convenient standard for proton and carbon-13 nmr spectra for a number of reasons. Not only are the twelve protons and four carbons highly shielded, but also the material is highly volatile, can be added in small amounts, is easily removed if the sample has to be recovered, and does not usually interact with samples.

The magnitudes of chemical shifts turn out to be of the order of parts per million ($10^{-6}$) of the operating frequency. For protons the usual range is about 10 ppm while for carbon-13 the full range is over 200 ppm. Other nuclei can have much larger ranges depending on their chemical environments. In practice we can define a chemical shift scale where chemical shift is given the symbol $\delta$, in terms of a ratio of frequency difference between the resonanace and that of TMS, and operating frequency. Eq. 1.3a is the exact definition of the $\delta$ scale.

$$\delta = \frac{\nu_R - \nu_{TMS} \times 10^6}{\nu_{spectrometer}} \qquad (1.3a)$$

where $\nu_R$ is the resonance frequency of the nucleus in question, $\nu_{TMS}$ is the resonance frequency of TMS, and $\nu_{spectrometer}$ is the operating frequency in MHz.

By definition TMS resonates at $\delta = 0$ (for protons and carbon-13) and almost all resonances appear to be left of TMS on the nmr spectrum.

From Eq. 1.3a you can see that $\delta$ is dimensionless as it is a ratio of the Larmor frequency shift with the spectrometer operating frequency. A typical resonance might occur 180 Hz downfield from that of TMS. If the operating frequency is 60 MHz then:

$$\delta = \frac{180 \text{ Hz}}{60 \times 10^6 \text{ Hz}} \times 10^6$$

$$= 3.0\ 10^{-6} \times 10^6$$

$$= 3.0 \text{ ppm}$$

Hence the frequency ratio of $3.0 \times 10^{-6}$ is multiplied by $10^6$ to give an easily handled number and is said to have units of parts per million, ppm.

Π A resonance has a chemical shift of $\delta = 3.0$ ppm at an operating frequency of 60 MHz. What is the frequency difference between that resonance and that of TMS?

We can use eq. 1.3a to calculate the answer:

$$\delta = 3.0 = \frac{\nu_R - \nu_{TMS}}{60 \times 10^6} \times 10^6$$

$$\nu_R - \nu_{TMS} = 60 \times 3.0 = 180 \text{ Hz}$$

So the frequency difference is 180 Hz.

We can think about chemical shifts either in terms of ppm or in terms of frequency depending on what we need to know. However, as the definition of $\delta$ involves a ratio of frequency difference to operating frequency and as the frequency difference increases proportionally with operating frequency, any chemical shift expressed in $\delta$ are *independent* of the operating frequency. This is just a rather long winded way of saying that if a resonance is at $\delta = 3.0$ ppm at an operating frequency of 60 MHz then it is still $\delta = 3.0$ ppm at any other operating frequency, 100 MHz, 220 MHz or whatever. Nevertheless, the *frequency difference* at other operating frequencies will be proportionally different.

Π What is the frequency difference between a resonance at $\delta$ = 3.0 ppm and that of TMS at an operating frequency of 500 MHz?

Eq. 1.3a gives:

$$\nu_R - \nu_{TMS} = 500 \times 3.0 = 1500 \text{ Hz}$$

Alternatively, if we know the frequency difference to be 180 Hz at 60 MHz it must be:

$$\frac{180}{60} \times 500 = 1500 \text{ Hz at } 500 \text{ MHz}$$

Finally in this section you should note that resonances to the left of TMS came from nuclei that are less shielded than those of TMS. The further to the left of the spectrum the greater the *deshielding*. Also, nuclei that are deshielded are said to resonate at *higher frequency*, or *to low field of*, or *downfield* from TMS. Fig. 1.3b summarises these terms.

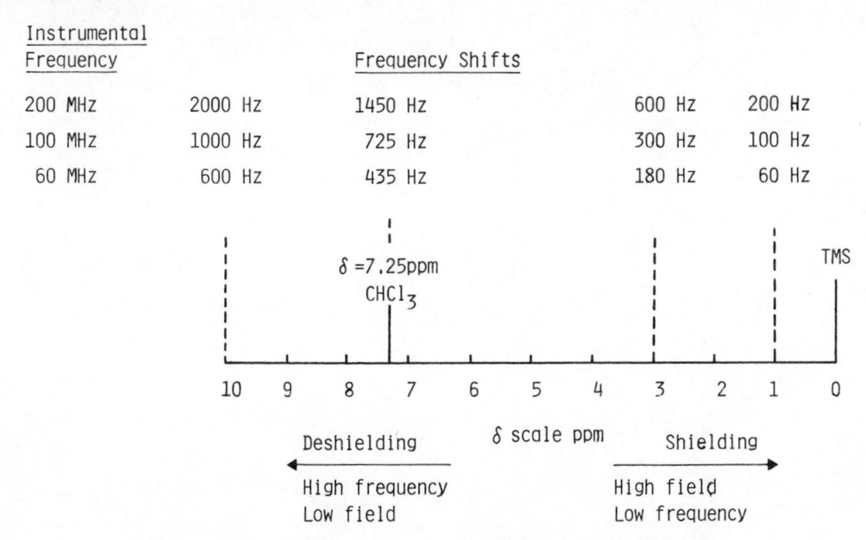


**Fig. 1.3b.** *Summary of chemical shift terms*

This section has contained a lot of ideas. Try the self-assessment questions, but don't worry if you have not picked up all the points yet. You can always go over this material again.

**SAQ 1.3a**

Indicate by circling T for true or F for false which of the following statements concerning chemical shift are correct.

(*i*) The chemical shift of a nucleus is defined as the shift in ppm from the resonance frequency of a standard, usually TMS.

T / F

(*ii*) The chemical shift of TMS is always 0 ppm.

T / F

(*iii*) The chemical shift of a nucleus will always be to the left of the reference in an nmr spectrum.

T / F

(*iv*) When looking at an nmr spectrum the resonance frequency increases on going from left to right.

T / F

**SAQ 1.3b**

The proton in chloroform ($CHCl_3$) is found to resonate at $\delta = 7.25$ ppm on a 60 MHz instrument.

Calculate:

(*i*) The frequency difference in Hz between the $CHCl_3$ resonance and that of TMS.

(*ii*) The frequency difference between the $CHCl_3$ and TMS resonances if the operating frequency were (*a*) 100 MHz, (*b*) 220 MHz.

(*iii*) The chemical shift, $\delta$, of the $CHCl_3$ resonance at (*a*) 100 MHz, (*b*) 220 MHz.

**SAQ 1.3c**

A forgetful nmr operator omitted TMS from a sample for nmr analysis. The sample was dissolved in $CDCl_3$ and the operator correctly positioned the resonance of residual $CHCl_3$ at $\delta = 7.25$ ppm. Two further peaks were observed in the sample, one 2 ppm downfield from the $CHCl_3$ resonance and one 3.5 ppm upfield from it. What were the chemical shifts of these two resonances?

## 1.4. FACTORS AFFECTING CHEMICAL SHIFT

While the subject of relating chemical shifts to structural features (functional groups) is going to be studied in Part 3, Qualitative Analysis, this seems an appropriate point to have a look at some factors which influence the shielding around protons and carbon-13 and so affect their chemical shifts.

We shall consider three general factors:

substitution and hybridisation;

inductive and mesomeric effects;

hydrogen-bonding.

In proton nmr *substitution* of a proton by carbon causes deshielding, ie the remaining protons resonate at lower field,

| eg | $R—CH_3$ | $R—CH_2—R$ | $(R)_2CHR$ | (R = alkyl) |
|---|---|---|---|---|
| proton resonance: ($\delta$ ppm) | 0.95 | 1.2 | 1.5 | |

Whether a carbon-13 is $sp^3$, $sp^2$, or sp *hybridised* influences the resonance position, as shown in Fig. 1.4a

| | $\delta_H$ (ppm) C—H | $\delta_C$ (ppm) C—H |
|---|---|---|
| Alkyl ($sp^3$) | 1–4 | 10–50 |
| Alkenic ($sp^2$) | 5–7 | 100–150 |
| Aromatic ($sp^2$) | 7–9 | 100–150 |
| Carbonyl ($sp^2$) | 8–10 | 150–200 |
| Alkynic (sp) | 3 | 50–80 |

**Fig. 1.4a.** *Effect of hybridisation on chemical shift*

To explain why there should be these variations in chemical shifts for differently hybridised carbon and hydrogen nuclei we have to introduce the idea of *anisotopy* in chemical bonds. This term means that the electron density in bonds is not the same in all directions. If the electron density is different in different directions then the shielding associated with the electrons is also going to vary around about the bond. Fig. 1.4b shows the anisotropic effect associated with $\pi$ bonds.

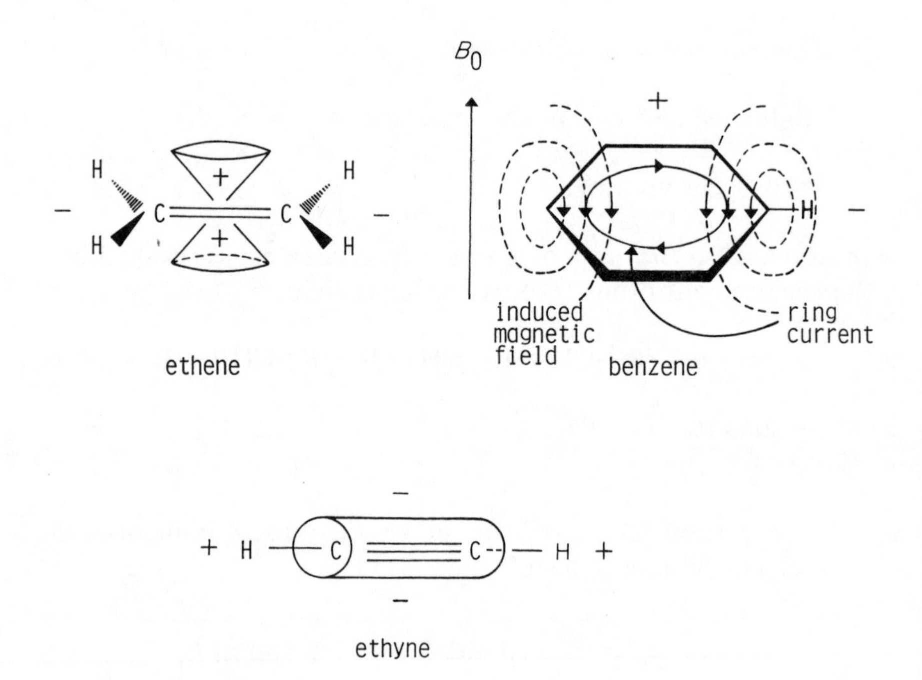


**Fig. 1.4b.** *Anisotropy of shielding in $\pi$-bonded systems*

The presence of the $\pi$-bond found in $sp^2$ hybridised systems, such as alkenes and carbonyls, gives rise to a shielding cone above and below the plane of the bond while those nuclei (protons and carbon-13) that lie in the plane of the bond are highly *deshielded* and so have chemical shifts well downfield of nuclei in fully saturated bonds.

The effect is even more pronounced for aromatic protons because the $\pi$ electrons of the aromatic ring when in the external magetic field, $B_O$, circulate around the ring. This effect, known as a ring current, produces a secondary (or induced) magnetic field which reinforces the applied field in the region of the aromatic protons (Fig. 1.4b). So the protons are highly deshielded.

For alkynic protons (sp hybridised) the anisotropic effect is such as to shield nuclei lying along the axis of the triple bond (Fig. 1.4b), and for carbonyl carbon atoms and aldehydic protons there is not only the anisotropy of the $\pi$-bonds to consider, but also *inductive* and *mesomeric* effects as outlined next.

Electronic effects which alter the polarity of bonds cause chemical shifts to vary; by and large any electron-withdrawing effect leads to deshielding, as shown in Fig. 1.4c. In the methyl halides the increasing electronegativity of the halogen, from iodine to fluorine causes an increase in polarity of the carbon-halogen bond, electrons being withdrawn from the carbon.

| Structure | $\delta_H$ | $\delta_C$ ppm |
|---|---|---|
| $CH_3$—I | 2.3 | −20.3 |
| $CH_3$—Br | 2.7 | 8.9 |
| $CH_3$—Cl | 3.0 | 23.8 |
| $CH_3$—F | 4.5 | 74.1 |
| —CHO | 8–10 | 200* |
| —$CONH_2$ | variable | 172* |
| —COCH=$CH_2$ | 5–7 | 194* |

*carbonyl carbon

**Fig. 1.4c.** *Effect of inductive and mesomeric effects on chemical shift*

Thus both the carbon-13 and proton resonances are deshielded (Fig. 1.4d). Similarly, the mesomeric effect influences the position of the carbonyl carbon-13 resonance from the 'standard' aldehydic value of $\delta_C$ = 200 ppm to the more shielded values of the unsaturated carbonyl and amide. Fig. 1.4d shows how the canonical forms reduce the slight positive charge associated with the carbonyl carbon.

i) $\overset{\delta+}{CH_3}-\overset{\delta-}{X}$ electronegativity inducing electron withdrawal

ii) aldehyde C=O with $\delta-$ on O and $\delta+$ on C, bonded to H: normal carbonyl polarisation

amide $-C(=O)-NH_2$ $\longleftrightarrow$ $-C(-O^-)=\overset{+}{N}H_2$ mesomeric diminution of $\delta+$ on the carbonyl carbon

$-C(=O)-CH=CH_2$ $\longleftrightarrow$ $-C(-O^-)=CH-\overset{+}{C}H_2$

**Fig. 1.4d.** *Inductive and mesomeric effects on structures*

The third general factor influencing chemical shifts, *hydrogen-bonding*, causes hydroxyl resonances in alcohols and carboxylic acids, and amino proton resonances in amines and amides to vary enormously. Fig. 1.4e gives some typical values, but all of these chemical shifts are prone to large concentration and temperature dependent variations.

| | | $\delta_H$ ppm |
|---|---|---|
| Carboxylic Acid Protons | $-CO_2H$ | 10–13 |
| Alcohol Hydroxyl Protons | —OH | 3–7 |
| Amine, Amide Protons | —NH | 2–5 |

**Fig. 1.4e.** *Typical chemical shifts for hydrogen-bonded protons*

Finally, in this section, I must stress that there are other factors which affect chemical shifts, and that I have deliberately not offered too much detailed explanation of how the factors above actually change chemical shifts. Nevertheless, these points should help you explain quite a few of the chemical shift variations that you are likely to come across.

**SAQ 1.4a**

Rank the spin-active nuclei in the following sets in their likely chemical shift order lowest to highest (ie most shielded to least). Note that precise chemical shifts are not required.

eg $-CH_2-$, $-CH_3$, $-CH$

Answer $-CH_3$ > $-CH_2-$ > $-CH$
for both proton and carbon-13.

(*i*) benzene ($C_6H_6$), methanal ($CH_2O$), ethyne ($HC{\equiv}CH$)

(*ii*) $CHCl_3$, $CH_2Cl_2$, $CCl_4$, $CH_3Cl$

(*iii*) Methanal ($CH_2O$), acetone ($CH_3COCH_3$), methyl formate ($CH_3OCOH$).

(*iv*) the different protons only in salicylic acid $C_6H_4(OH)(CO_2H)$ (treat the aromatic protons as being just one type):

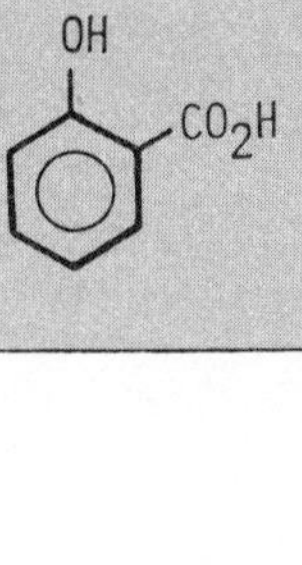

OH
$CO_2H$

→

**SAQ 1.4a**

**SAQ 1.4b**

Which of the following statements about anisotropy in chemical bonds are correct? Circle T (true) or F (false).

(*i*) Anisotropy in chemical bonds means the electron density and hence the shielding effect of bonds is different in different regions around the bond.

T / F

(*ii*) Anisotropy always leads to spin-active nuclei being deshielded.

T / F

(*iii*) Aldehyde protons and carbonyl carbon atoms owe their chemical shifts solely to anisotropic effects

T / F

(*iv*) Anisotropy occurs only in $\pi$-bonds.

T / F

## 1.5. SPIN–SPIN COUPLING

In the last section you saw how the electrons in a structure could influence the precise position of resonance of a spin-active nucleus. If we now look at an nmr spectrum, the proton spectrum of diethyl ether ($CH_3CH_2OCH_2CH_3$), Fig. 1.5a, we see not the two expected resonances for the two chemically different types of proton, but two *sets* of resonances, one set being a quartet centred at $\delta$ = 3.4 ppm and the other a triplet centred at $\delta$ = 1.2 ppm. Now the quartet turns out to be the resonance for the $CH_2$ protons and the triplet is the resonance for the $CH_3$ protons, but why are they split into these symmetrical patterns?

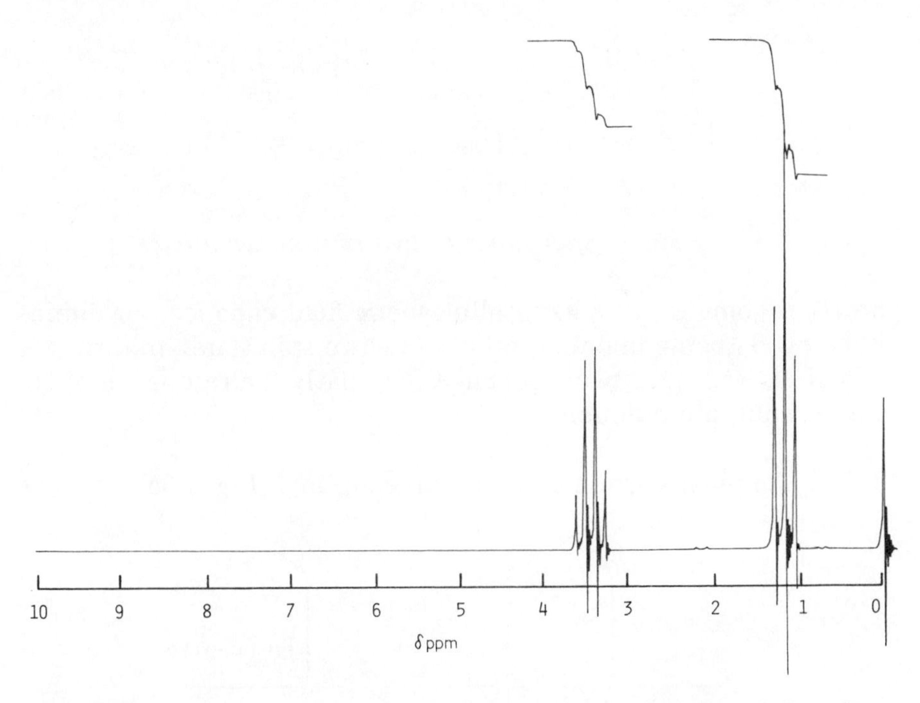


**Fig. 1.5a.** *Proton spectrum of diethyl ether* $CH_3CH_2OCH_2CH_3$

To start the answer let us look at a simpler system where there are only two spin-active nuclei, protons, on adjacent carbon atoms (Fig. 1.5b). We will call these nuclei $H_A$ and $H_X$ and assume that they have different chemical shifts (as for example in $Br_2CHCHCl_2$). Now, as well as the external magnetic field, $B_O$, that these protons feel and as well as the perturbing magnetic fields of the electrons which cause the protons to have different chemical shifts, each of these two nuclei experience a slightly different magnetic field depending on the spin state of the neighbouring nucleus. That may sound puzzling, but as shown in Fig. 1.5b, $H_A$ could have one of two magnetic spin states ($\alpha$ or $\beta$) depending on whether it is aligned against or with the external field $B_O$; similarly for $H_X$.

$>C-C<$ (each C bonded to H: $H_A$, $H_X$)

| | $H_A$ | $H_X$ |
|---|---|---|
| Spin | $\alpha$ | $\alpha$ - lined up against $B_0$ - high energy |
| States | $\beta$ | $\beta$ - lined up with $B_0$ - low energy |

**Fig. 1.5b.** *Spin states of two spin-active nuclei*

So $H_A$ is going to have its local magnetic field enhanced or diminished by $H_X$ being in one or other of its two spin states; in turn that means $H_A$ will have two resonances. Similarly the resonance of $H_X$ will be split into a doublet.

This is the phenomenon of spin–spin coupling. (Fig. 1.5c).

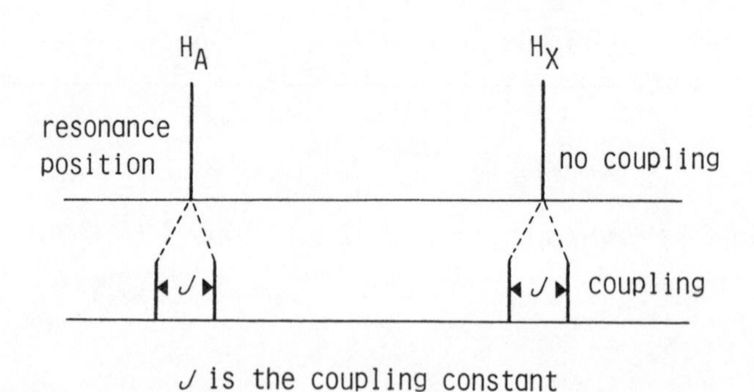


**Fig. 1.5c.** *Doublets from coupling of two spin-active nuclei*

Usually, there would be about equal numbers of protons in the two spin states (as a consequence of the Boltzmann distribution, eq. 1.1c) so that there is a more or less equal chance of $H_A$ 'seeing' $H_X$ in the $\alpha$ or $\beta$ spin state and thus the splitting pattern is a doublet with both signals of equal intensity; similarly for $H_X$.

The splitting between the two peaks of each doublet is the same, and is said to be the coupling constant, $J_{AX}$. (The subscript AX refers to the two nuclei which are coupled). Unlike chemical shifts, $J_{AX}$ is measured in hertz, and also unlike chemical shifts, $J_{AX}$ is independant of the size of the external field $B_O$ or operating frequency. This means that for a given coupled system $J_{AX}$ is the same whether the spectrum is recorded at 60, 220, or 500 MHz.

The magnitude of $J$ can be viewed as a measure of the efficiency of the coupling interation between neighbouring spins. However, this is a complex topic and beyond the scope of this Unit.

Now what would happen if we had a spin system containing three nuclei, two of which had the same chemical shift and one of which was different? An example of such a system could be $ClCH_2CHO$ (chloroacetaldehyde). The $CH_2$ protons would 'see' the aldehyde proton in two possible spin states and so be split into a 1 : 1 doublet as in the last example. But the aldehyde proton will 'see' two protons in a number of combinations of spin states (Fig. 1.5d). Both the $CH_2$ protons could be in a $\alpha$ spin states ($\alpha\alpha$), or $\beta$ spin states ($\beta\beta$) or one could be $\alpha$ and the other $\beta$ ($\alpha\beta$), or vice versa ($\beta\alpha$). So there are four possible spin state combinations although both $\alpha\beta$ and $\beta\alpha$ spin states are equal in energy. So the aldehyde proton 'sees' three energy states for the $CH_2$ protons with a 1 : 2 : 1 probability and thus resonates with a fine structure of a 1 : 2 : 1 triplet. As before the spacing between the lines in the splitting pattern is the coupling constant $J$.

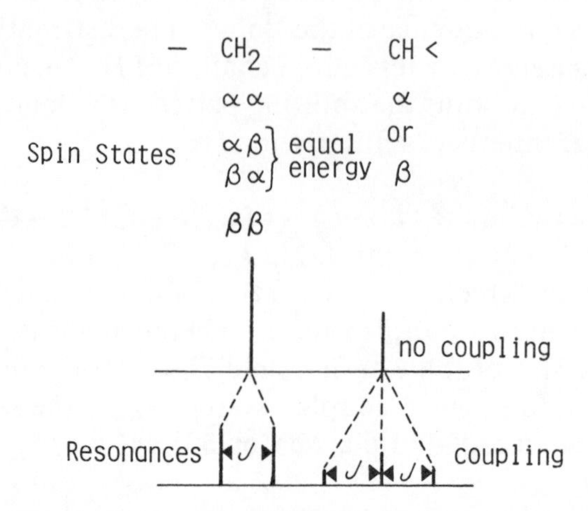


**Fig. 1.5d.** *Spin states and resonances for a $-CH_2-CH<$ system*

The general rule which is emerging from the above observations and explanations is a very important one for qualitative nmr spectroscopy. When a group of protons ($H_A$) is coupled to a group of protons ($H_X$) containing '$n$' equivalent protons then the resonance of the $H_A$ protons will be split into ($n$ + 1) peaks. In fact the rule is even more general than this and we will look at the most general form later, but for now with proton systems we can see that the splitting pattern gives us information about the number of protons in a group adjacent to the group whose resonance is split.

Let us now come back to the triplet and quartet of diethyl ether. Fig. 1.5e shows how these must arise. The argument for the triplet is the same as above, there are three energy states for the $CH_2$ protons so there will be a triplet for the adjacent $CH_3$ resonance ($n = 2$ so $n + 1 = 3$ peaks). For the $CH_3$ protons there are four energy states corresponding to the various combinations of spin states shown in the figure. So there must be four lines in the ratio 1:3:3:1 in the coupling pattern for the $CH_2$ resonance ($n = 3$ so $n + 1 = 4$ peaks).

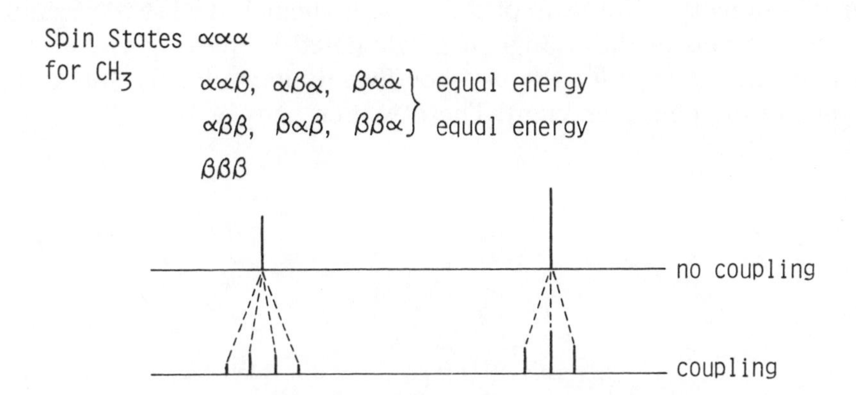


**Fig. 1.5e.** *Spin states for $CH_3$ protons and resonances of the $-CH_2CH_3$ system*

Π What splitting pattern might be expected for the protons in an isopropyl group, $(CH_3)_2CH-$?

Applying the $(n + 1)$ rule to the $CH_3$ groups first we can see that their resonance will be a doublet because they are adjacent to a single proton. The CH resonance will be a heptet ($n = 6$ so $n + 1 = 7$).

Before turning to coupling involving carbon-13 nuclei there are three further points about proton spin-coupling. First, the size of coupling constants is usually between 0 and 15 Hz (see section on qualitative analysis for examples). This means that if a set of peaks have spacings of greater than 15 Hz they are unlikely to be a spin pattern.

Secondly, the relative intensities of the peaks in a spin pattern can be worked out, as we have seen, by counting the number of equal energy spin states in the coupling nuclei. However, an easy method of remembering these intensities is provided by Pascal's Triangle (Fig. 1.5f). This mathematical device is used to calculate the coefficients of the binomial distribution, but these same coefficients are the relative intensities of the peaks in a splitting pattern. So if a resonance is split into five peaks—a quintet—then the coefficients are given by the fifth level of the triangle, namely 1 : 4 : 6 : 4 : 1. Subsequent levels of the triangle are calculated by starting a row with 1 and adding together the two coefficients immediately above the space of the new coefficient. The row then ends in 1.

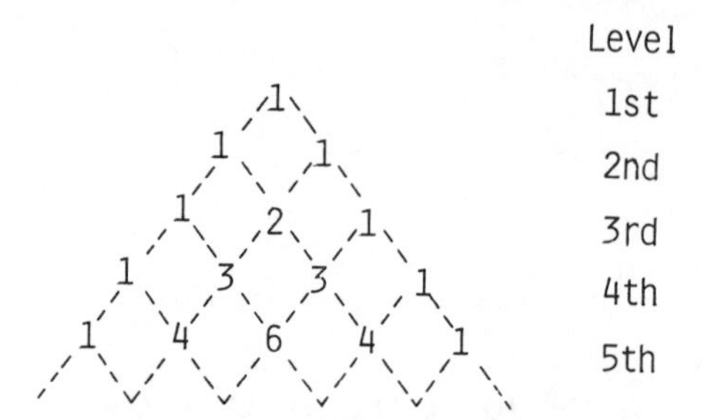


**Fig. 1.5f.** *Pascal's triangle*

Thirdly, we have seen how these symmetrical splitting or spin coupling patterns arise and how they provide structural information. Unfortunately, not all spin coupling leads to simple patterns. Sometimes, for reasons beyond the scope of this course, extremely complex, non-symmetrical fine structure occurs. (You will see some examples later). In such cases, called second-order spectra, a much more detailed theoretical treatment is required. Interested readers should consult the references given in the Bibliography.

**SAQ 1.5a**

Which of the following statements about spin–spin coupling are correct?

Indicate your answer by circling T (true) or F (false).

(*i*) Spin–spin coupling arises because one spin-active nucleus interacts with the spin states of nearby spin-active nuclei.

T / F

**SAQ 1.5a (cont.)**

(*ii*) Spin–spin coupling depends on the size of the external magnetic field $B_O$.

T / F

(*iii*) The coupling constant, $J$, is always measured in Hertz.

T / F

**SAQ 1.5b**

The spectrum of a sample sent for nmr analysis showed two peaks separated by 0.1 ppm at 60 MHz. If the two peaks are in fact a doublet arising from coupling which *one* of the following would be the splitting at 220 MHz?

(*i*) 0.1 ppm

(*ii*) 6 Hz

(*iii*) 22 Hz

**SAQ 1.5c**

In the compound dipropyl ether

$$(CH_3CH_2CH_2OCH_2CH_2CH_3)$$

the central methylene protons' resonance is a sextet. Which of the following is the likely intensity ratio for the six peaks?

(*i*) 1:2:3:3:2:1

(*ii*) 1:2:1:1:2:1

(*iii*) 1:7:9:9:7:1

(*iv*) 1:5:10:10:5:1

## 1.6. SPIN COUPLING IN CARBON-13 NMR

We have been discussing spin-coupling between nuclei of the same type—*homonuclear* spin coupling. However, coupling can take place between spin-active nuclei of different species—*heteronuclear* spin-coupling. In particular, coupling between protons and carbon-13 is rather important. The upper part of Fig. 1.6a shows what is known as the 'off-resonance decoupled' carbon-13 spectrum of diethyl ether. This term means that all the proton carbon-13 couplings have been destroyed except for those between a carbon nucleus and protons directly bonded to it. So we see a quartet at $\delta_C$ = 14.1 ppm and this, by the ($n$ + 1) rule, must mean that the carbon is bonded to three protons, ie the resonance belongs to the $-CH_3$ group. Similarly, the triplet at $\delta_C$ = 64.8 ppm must belong to the $-CH_2-$ group.

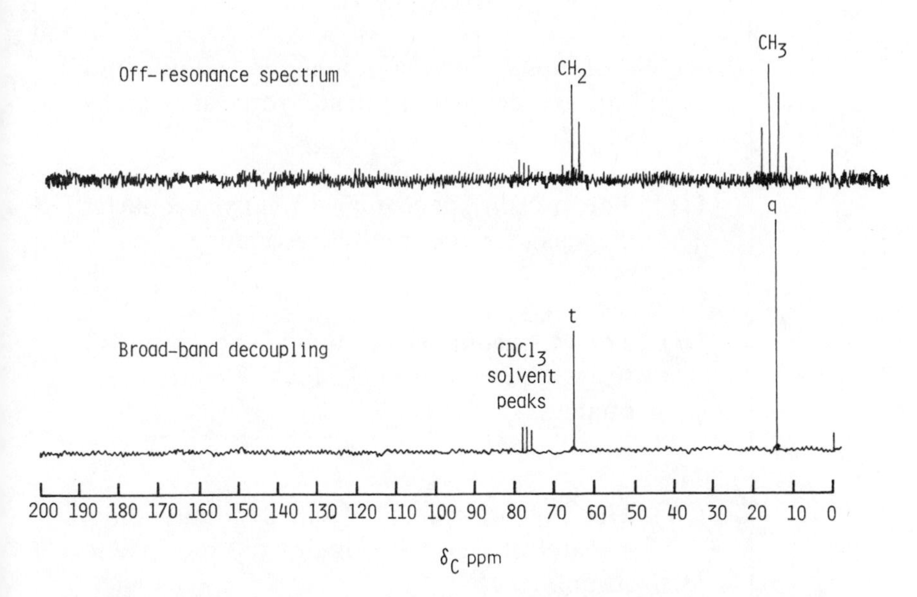


**Fig. 1.6a.** *Carbon-13 spectra of diethyl ether in* $CDCl_3$

There are two important differences between observed couplings in carbon-13 spectra and proton spectra. For instrumental reasons the pattern of peaks is often highly distorted and the spacings between the splittings is usually not the true coupling constant. So we use these off-resonance spectra as a qualitative guide as to how many protons are bonded to a given carbon atom.

There is a second type of carbon-13 spectrum in which all the couplings between protons and carbon have been destroyed. This type is known as a 'broad-band decoupled' spectrum and is used for assigning precise chemical shifts. The lower part of Fig. 1.6a shows this. All carbon-13 spectra from now on in this Unit will be displayed as broad-band decoupled spectra with the off-resonance information shown as a letter above each peak (q – quartet, t – triplet, d – doublet, s – singlet).

**SAQ 1.6a**

Concerning the molecular fragment

$$CH_3CH_2CH_2-,$$

which of the following statements about spin coupling are correct. Indicate your answer by circling T (true) or F (false).

(*i*) The proton spectrum will show a quartet of peaks for the methyl resonance.

T / F

(*ii*) The off-resonance carbon-13 spectrum will show a quartet of peaks for the methyl resonance.

T / F

(*iii*) The proton spectrum will probably show a sextet of peaks for one of the methylene resonances.

T / F

⟶

**SAQ 1.6a (cont.)**

(*iv*) The off-resonance carbon-13 spectrum for the methylene carbons will show two triplet resonances with ratios of intensities 1 : 2 : 1.

T / F

## 1.7. THE $(2nI + 1)$ RULE

Earlier I introduced the $(n + 1)$ rule to help predict the number of peaks in a spin-coupled pattern involving protons. In fact this rule is a simple form of a more general case—the $(2nI + 1)$ rule—that can be used to produce the number of lines in any simple homonuclear and heteronuclear spin pattern. In the formula '$n$' stands for the number of equivalent nuclei coupling to the nucleus under consideration, while $I$ is their spin quantum number. For protons, carbon-13 and fluorine-19 $I = \frac{1}{2}$ and so the formula reduces to $(n + 1)$. But most other nuclei have different values for $I$. In fact if $I = 0$ the nucleus is not spin active, eg carbon-12.

This rule helps explain the presence of some patterns of peaks in carbon-13 and proton nmr spectra that have nothing to do with the compound being studied. For example, in the off-resonance carbon-13 spectrum of diethyl ether (Fig. 1.6a) there is a 1:1:1 triplet of peaks at $\delta_C = 77.0$ ppm. This resonance is the carbon-13 of the solvent, $CDCl_3$, coupled to the deuterium. As deuterium has a spin of 1 so the $(2nI + 1)$ rule gives $2 \times 1 \times 1 + 1 = 3$ peaks for the resonance of the carbon-13 nucleus.

The rule does *not* help in working out the relative intensities of peaks in a pattern and Pascal's Triangle holds only for spin-$\frac{1}{2}$ nuclei. Actually you have to work out the number and relative energies of all the spin states in the system. However, don't worry, we won't be pursuing relative intensities of patterns from nuclei with spins greater than $\frac{1}{2}$ in this Unit.

**SAQ 1.7a**

Which of the following will be the correct number of lines observed for the methyl carbon resonance in the carbon-13 spectrum of hexadeuteroacetone ($CD_3COCD_3$)?

(*i*) 4

(*ii*) 6

(*iii*) 7

(*iv*) 13

## 1.8. RELAXATION TIMES

The third nmr phenomenon, relaxation, which affects line widths and intensities, is often not discussed in introductory books dealing with continuous wave (CW) nmr, as meaningful data is difficult to obtain. However, with the increasing use of Fourier Transform (FT) nmr an elementary understanding of relaxation is essential. (Fourier Transform nmr is dealt with in Part 4).

If we think about an individual spin-active nucleus then in a magnetic field we know that it can be excited from its lower spin state to a higher spin state by radiofrequency radiation at its Larmor frequency. This is the phenomenon of resonance. However, having been excited the spin-active nucleus must lose that extra energy if it is to return to the lower spin state. The means whereby this energy is lost are known as relaxation. Fig. 1.8a shows this point.

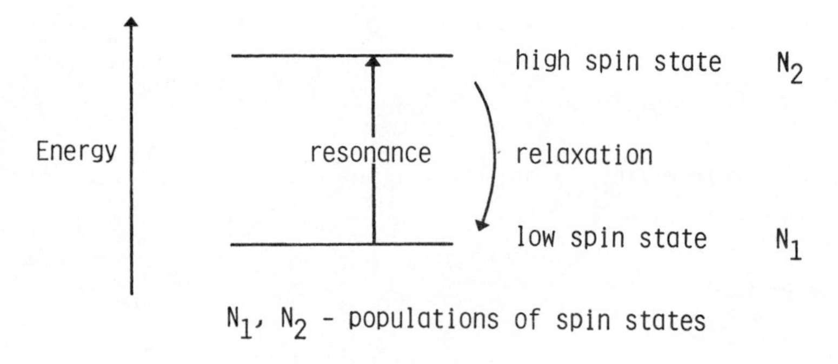


**Fig. 1.8a.** *Resonance and relaxation*

Considering not an isolated nucleus but a macroscopic number of nuclei in a magnetic field then they will distribute themselves among the various energy levels according to the Boltzmann Distribution. You probably remember the equation from Section 1.1.

$$N_2/N_1 = 1 - \Delta E/kT \qquad (1.1c)$$

where $N_2$ and $N_1$ are the populations of the higher and lower spin states, $\Delta E$ is the energy difference between the spin states, $k$ is the Boltzmann constant, and $T$ is the absolute temperature of the system.

If we were to substitute appropriate values into this equation it would tell us that the populations of the two spin states, when in equilibrium, are nearly equal; typically there is a difference of only about five spins per million excess in the lower energy state. This is one reason why the nmr experiment is inherently insensitive, (see Part 4), but it also shows us why relaxation is so important to nmr spectroscopy. If excited nuclei could not relax back to the low energy state then we would lose the nmr signal after just one scan in CW spectroscopy or one pulse in FT spectroscopy.

So how do nuclei relax? There are several ways, but we shall consider only two. In liquid samples the molecules are jostling around, spinning, oscillating, and randomly moving through the liquid. This process, as you probably know, is called Brownian motion. As there are magnetic fields associated with the nuclei and electrons in a molecule we can imagine Brownian motion causing rapid random changes in these molecular magnetic fields, a kind of 'magnetic storm'. At some time the local magnetic field near an excited spin nucleus is going to be just what is required to induce resonance, but this time the nucleus will move from high to low spin state. The excess energy has then been lost to the rest of the molecular motions. This type of relaxation is known as *spin–lattice* relaxation, the word lattice referring to the molecular world at large, surrounding the spin-active nucleus.

In a large sample of nuclei not all nuclei will relax at the same instant and so the high spin state *decays* over a period of time. This decay usually obeys a familiar type of rate equation—that of exponential decay (Eq. 1.8a).

$$\text{Rate of decay} = \text{fn}\ \exp(-1/T_1) \qquad (1.8a)$$

The true equation is much more complex and need not concern us. However, the parameter $T_1$ is a constant for a given structural type of spin-active nucleus and is known as the spin–lattice relaxation time. $T_1$ values can vary over a considerable range, but for typical organic materials in solution $T_1$ is usually between 0.5 and 50 seconds.

The second mechanism of relaxation is called *spin–spin relaxation* and, very simply put, it involves an excited spin state exchanging spins with an adjacent spin-active nucleus. The rate of relaxation by this means is again an exponential function like Eq. 1.8a, but is characterised by the spin–spin relaxation time, $T_2$. It can be shown that $T_2$ must be less than or equal to $T_1$, but for many practical purposes we can consider $T_1 = T_2$ and that there is just a characteristic relaxation time for a given nucleus bonded in a given way.

Having looked at some of the ideas behind relaxation in nmr we can now consider a couple of consequences. First, in proton nmr a very approximate value for the spin–spin relaxation time of a resonance is given by the inverse of the peak width at half height, Eq. 1.8b.

$$\text{Peak width at half height} = 1/T_2 \qquad (1.8b)$$

where $T_2$ is the spin–spin relaxation time

A typical singlet methyl resonance might have a peak width of 0.5 Hz at half height so the rough relaxation time would be about 2 seconds. Now while different types of proton will have different relaxation times which can be correlated with structural features, what is of concern in this example is a much more general point. If the relaxation time were to be, say, ten times faster then, by Eq. 1.8b, the peak width at half height would be 1/0.2, ie 5 Hz. So decreasing relaxation times leads to broad peaks. Conversely, long relaxation times will give very sharp peaks, but they will be difficult to observe because resonance will tend to equalise the populations of the spin states and with slow relaxation the nmr signal will vanish.

Broad signals are often observed in proton nmr spectra associated with O—H, and N—H resonances. The relaxation times of these types of proton are often shortened by chemical exchange processes and other nuclear phenomena (Fig. 1.8b).

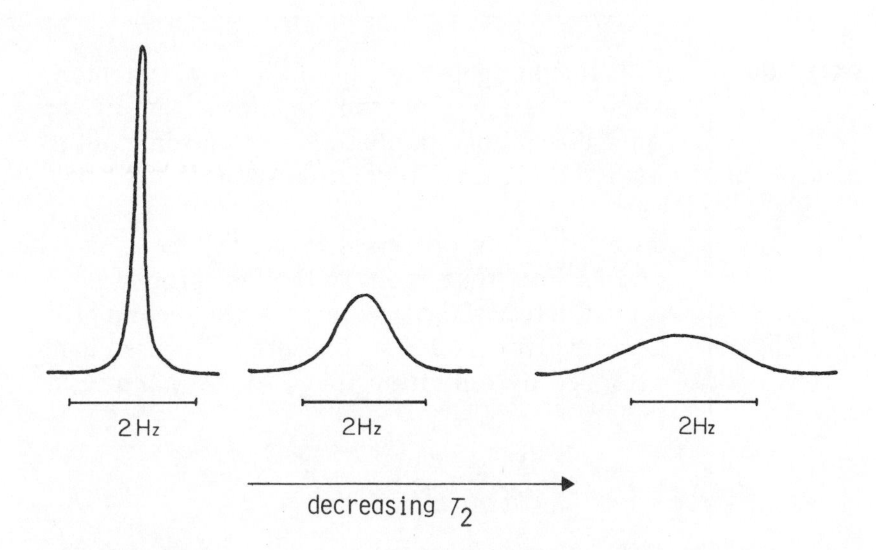


**Fig. 1.8b.** *Effect of relaxation time in signal width*

A second consequence of relaxation can be seen in carbon-13 spectroscopy where the signal intensity varies depending on how many hydrogen atoms are attached to the carbon atom. Methyl carbon atoms (3H) can give stronger signals than methylene carbons (2H) which in turn usually give stronger signals than methine carbons (1H). If there are no hydrogens attached as with quaternary carbons or carbonyl carbons (excluding aldehydes) then the carbon-13 resonances signal is often very weak. This variation of intensities is partly determined by instrumental conditions, but the physical origin is a relaxation effect. The second relaxation process involving spin exchange is important for carbon-13 relaxation. Carbon atoms form the skeleton of molecular structure and so are 'buried' in the molecular framework. Thus they are less susceptible to the fluctuating magnetic fields that allow spin–lattice relaxation. Yet for spin–spin relaxation the carbon-13 nucleus requires a spin-active nucleus adjacent to it. There is only a tiny chance that an adjacent carbon atom will be spin-active (remember carbon-12 is not spin-active and accounts for 98.9% of all carbon atoms) so the carbon-13 nucleus relies on protons for spin exchange. As the number of bonded protons decreases so does the efficiency of spin–spin relaxation and so the signal intensity goes down. Naturally, for carbons with no hydrogens attached the signal intensity is lowest.

**SAQ 1.8a**

Fill in the gaps in the following statement about relaxation phenomena from the list of phrases/symbols supplied. Note that some of the list are quite inappropriate responses.

Two relaxation processes for nuclei are ... 1 ... and ... 2 ... relaxation. In the first process ... 3 ... to the rest of the system while the second process involves ... 4 .... The processes are characterised by relaxation times ... 5 ... and ... 6 ....

(*i*) energy is lost by

(*ii*) energy is lost to

(*iii*) spin–lattice

(*iv*) $T_1$

(*v*) spin inversion

(*vi*) $T_{\frac{1}{2}}$

(*vii*) spin–spin

(*viii*) $T_2$

(*ix*) exchange of spins

**SAQ 1.8b**

Which of the following statements about relaxation phenomena are true?

Indicate your answer by circling T (true) or F (false).

(*i*) The relaxation time for a spin-active nucleus is the time taken for the nucleus to change from a high energy to low energy spin state.

T / F

(*ii*) The relaxation time for a spin-active nucleus depends on the type of nucleus.

T / F

(*iii*) Spin–spin relaxation is slower than spin–lattice relaxation.

T / F

(*iv*) Spin–lattice relaxation times depend on the viscosity of the sample.

T / F

**SAQ 1.8c**

A peak width at half height in an nmr resonance is found to be 0.3 Hz. Which of the following values is an approximate relaxation time for the resonance?

(*i*) 0.3 s

(*ii*) 3 s

(*iii*) 33 s

(*iv*) 0.03 s

## 1.9. INTEGRATION

In contrast to the previous nmr parameters which can be difficult to understand the idea behind the fourth nmr parameter, integration ratios, is very simple. The area under nmr signals is proportional to the number of spin-active nuclei causing the signal. So, if we have two or more separate resonances and if we measure the areas under each resonance signal we obtain the relative number of spin-active nuclei for each resonance.

The areas under the resonance signals are usually obtained by integration of the peaks. This can be done electronically by the instrument and the result displayed either as a 'step' curve, or integration curve, where the height of each step is proportional to the area underneath, or as a digital readout. Fig. 1.9 shows a simple example,

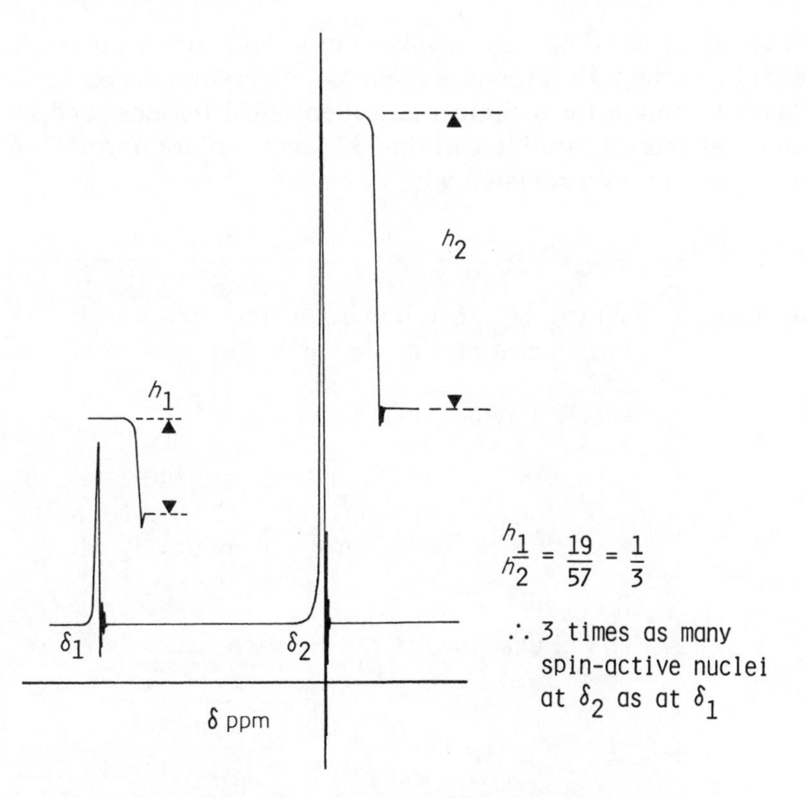


**Fig. 1.9.** *Step curve integration*

and Fig. 1.5a, the spectrum of diethyl ether also has the integration curve drawn. Careful measurement of the heights involved give a ratio of 2 : 3 for the areas and hence the relative number of protons associated with each resonance.

There are three points to stress concerning integrations. First, for most qualitative and quantitative purposes we measure the relative number of each kind of spin-active nucleus. Absolute numbers are much more difficult to obtain, so for the diethyl ether example we cannot tell from the integration that the true numbers of proton involved are 4 and 6, only that the ratio is 2 : 3.

Secondly, peak areas must be measured, not peak heights. If signals overlap then only the total area can be obtained accurately. Using peak areas, and with a properly adjusted spectrometer, an accuracy of measurement of about 5% is usually attainable.

Thirdly, the preceding notes apply to most spin-active nuclei, however, with carbon-13 resonances reliable integration curves are very difficult to obtain for technical and theoretical reasons such as relaxation effects. So routine carbon-13 spectra do not normally have integration curves associated with them.

**SAQ 1.9a**

Which of the following statements concerning integration of nmr signals is correct?

Circle T (true) or F (false).

(*i*) For a proton nmr spectrum the areas under each resonance are proportional to the number of protons causing the signal.

T / F

(*ii*) Peak heights can be used instead of peak areas.

T / F

**SAQ 1.9a (cont.)**

(*iii*) If a spectrum contains only one resonance there is little point in measuring its integration.

T / F

**SAQ 1.9b**

The diagram below shows a proton nmr spectrum complete with integration curve. What is the likely ratio of protons?

(*i*) 5 : 2 : 3

(*ii*) 8 : 4 : 6

(*iii*) 2 : 1 : 1.5

(*iv*) 4 : 2 : 3

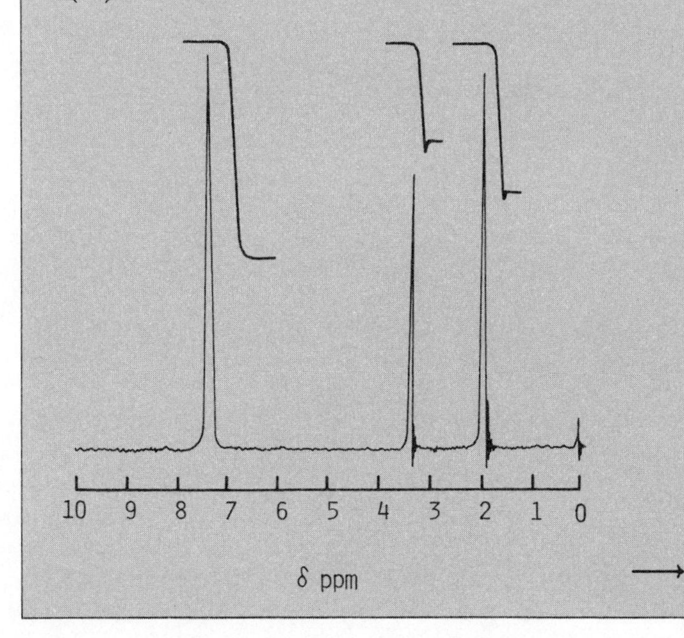


→

**SAQ 1.9b**

## 1.10. SUMMARY OF PART 1

We have covered a great deal of material in this first part of the Unit so it is worth summarising the main points. First, you learned about the origin of the nuclear magnetic resonance phenomenon; how spin-active nuclei like protons, carbon-13, and fluorine-19 can have different spin states when placed in a magnetic field. The energy difference between spin states corresponded to the energy of radiofrequency electromagnetic radiation and, if the correct Larmor frequency was applied to a spin-active nucleus, then it could change spin states. The fundamental equation governing this process was seen to be:

$$\nu = \frac{\gamma}{2\pi} B_O$$

where $B_O$ was the external magnetic field

and $\nu$ was the Larmor frequency.

With that idea in mind you saw how information could be extracted from an nmr spectrum in terms of four nmr parameters.

*Chemical shifts* measured in ppm from a standard such as TMS gave information about the chemical type of spin-active nucleus, eg whether protons were aliphatic, alkenic or aromatic.

*Coupling constants* measured in Hertz gave information about the number of adjacent spin-active nuclei via the $2nI + 1$ rule. The origin of spin coupling involved a spin-active nucleus being able to sense the various spin states of adjacent nuclei.

*Relaxation times* affected the intensity and shape of nmr signals and involved loss of spin energy to the molecular system—spin-lattice relaxation—or transfer of spins between nuclei—spin–spin relaxation. The rates of these processes were noted to be exponential decays governed by time constants $T_1$ and $T_2$.

*Integrations* or integration ratios of nmr signals gave information about the relative number of spin-active nuclei associated with each resonance as the area under a signal was proportional to the number of spins causing the signal.

With all this potential information available you can perhaps see why nmr spectroscopy is such an important instrumental, analytical technique. In the next Parts we will see how this information is obtained and used.

**SAQ 1.10a**

Which of the four nmr parameters, chemical shifts, spin–spin coupling constants, relaxation times and integrations could give you information on:

(*i*) the number of spin-active nuclei next to a resonating nucleus;

(*ii*) chemical exchange processes;

(*iii*) functional groups;

(*iv*) the relative amounts of the components of a mixture.

Now that you have arrived at the end of Part 1 you should be able to achieve most of the following 40 or more objectives. This is some achievement! Don't worry if you are not perfect. I suggest you read through the list, note those of which you are unsure, and go back and check them out.

**Objectives of Part 1**

Now that you have completed this Part you should be able to accomplish many of the objectives given below. We have covered a lot of ground and for most students some additional work may be needed if all the objectives are to be met. Go through the list below and note what areas need some further study. You should then be able to:

- explain the origin of the nmr phenomenon;

- recognise that spin-active nuclei can have energetically different spin states when placed in a magnetic field;

- label spin states as $\alpha$ or $\beta$ depending on whether they are high or low energy states;

- appreciate that radiofrequency induced transitions between spin states is the basis of nuclear magnetic resonance;

- draw an energy description of the nmr phenonmenon;

- state the relationship of the Larmor resonance frequency of a spin-active nucleus to the magnetic field strength as

$$\nu = \frac{\gamma}{2\pi} B_0;$$

- recognise, via Boltzmann's Distribution, that there is very little difference in the populations of spin states;

- appreciate that spin-active nuclei have different Larmor frequencies;

- state that spin-active nuclei are those which have a non-zero value for the spin quantum number ($I$);

- appreciate that the spin quantum number determines the number of spin states that a nucleus may have via the expression ($2I + 1$);

- state that there are four nmr parameters, chemical shift, spin–spin coupling, relaxation time, and integration that can give qualitative and quantitative analytical information;
- explain the basis of the chemical shift parameter;
- describe the idea of electronic shielding of nuclei;
- define chemical shift as

$$\delta = \frac{\nu_R - \nu_{TMS}}{\nu_{spectrometer}} \times 10^6 \text{ ppm}$$

- appreciate the use of tetramethylsilane (TMS) as a reference material;
- calculate chemical shifts in terms of hertz when given in ppm ($\delta$) and vice versa;
- recognise the apparent constancy of chemical shifts at different operating frequencies as a consequence of the definition of chemical shifts;
- describe how three general factors, substitution/hybridisation, inductive/mesomeric effects, and hydrogen bonding can influence the chemical shift of a spin-active nucleus;
- explain how anisotropy of chemical bonds can influence chemical shifts;
- apply the structural considerations of the previous two objectives to predict approximate chemical shifts for proton and carbon-13 nuclei in simple organic structures;
- describe the basis of spin–spin coupling;
- recognise that spin-active nuclei can sense the spin state of neighbouring nuclei and so have their local magnetic fields altered by neighbouring spin states;

- predict simple splitting (coupling) patterns for coupled nuclei;
- calculate the number of lines in a simple splitting pattern from the $(2nI + 1)$ rule;
- calculate the relative intensities of the peaks in a simple splitting pattern from the number of spin states and, for spin $\frac{1}{2}$ nuclei, from Pascal's Triangle;
- measure spin coupling constants of simple patterns from an nmr spectrum;
- appreciate that coupling constants, $J$, are measured in hertz and are independent of the spectrometer's operating frequency;
- recognise that some spin coupled systems give spectra too complex for easy interpretation (second order systems);
- describe aspects of spin coupling in carbon-13 systems;
- explain the terms 'off-resonance' and 'broad-band decoupled' when applied to carbon-13 spectra;
- appreciate that while coupling patterns are observed between protons and carbon-13 nuclei, reliable coupling constants are difficult to obtain;
- predict off-resonance coupling patterns for carbon atoms with zero, one, two, or three protons attached;
- describe the basis of relaxation in nmr spectroscopy;
- recognise the importance of Boltzmann's Distribution in nmr relaxation;
- appreciate the relaxation mechanisms are important in nmr spectroscopy to maintain a population difference between spin states;
- draw an energy description of resonance and relaxation;

- describe two mechanisms of relaxation, spin–lattice and spin–spin, in terms of energy dissipation to the surrounding lattice and exchange of spin states respectively;
- describe a general exponential equation that governs the rate of relaxation;
- recognise that relaxation times are constants in a rate equation;
- describe the effect upon peak widths of long and short relaxation times;
- state that the inverse of peak width of an nmr signal at half height is a rough guide to relaxation time for that nucleus;
- describe the importance of relaxation phenomena in determining signal intensities in carbon-13 nmr;
- describe the basis of integration of nmr signals;
- appreciate that the area under an nmr signal is proportional to the number of nuclei causing the signal
- measure integrations from step curves on nmr spectra;
- appreciate the use of integration ratios;
- recognise that peak heights do not give a true reflection of the number of nuclei associated with a resonance;
- state that carbon-13 spectra do not usually give reliable integrations for technical reasons;
- appreciate the types of structural information available via the nmr parameters, viz: chemical type of spin-active nucleus from chemical shifts; number of adjacent nuclei via spin coupling and the $(2nI + 1)$ rule; the relative amounts of spin-active nuclei from integrations.

# 2. Sample Handling and Instrumentation

### Overview

Now you have a lot of nmr ideas and jargon under your belt and if you are not in too much of a spin (a prize is offered for the worst nmr pun) we had better find out about how to prepare a sample for nmr analysis and what an nmr spectrometer is.

In this Part we will first look at some of the factors in sample preparation for nmr analysis, and then at various instrumental features common to all nmr spectrometers from the routine instrument to the sophisticated research spectrometer. We consider these two experimental aspects, sample preparation and instrumentation, together because the nmr spectrometer is the means whereby the information contained in the sample is extracted. The interpretation of the information is treated in Parts 3, 4 and 5, but the information obtained is going to depend on the sample and spectrometer.

We shall look at nmr solvents and see that they are preferably aprotic, often deuterated, material. We shall consider the amounts of sample required in terms of the concentrations of spin-active nuclei, and we will encounter the phenomenon of deuterium exchange, a useful technique for spotting labile protons.

By the end of this part you should be able to describe the main features of an nmr spectrometer, viz, a magnet, a radiofrequency source, a probe, a detector system and a data processing system. You will also have found out about homogeneity, spinning side bands, FT and CW instruments.

Don't be alarmed if most of these terms are unfamiliar because you are going to learn about them now.

## 2.1. SAMPLE PREPARATION

We are concerned here only with solution or liquid phase samples. Solid state and gas phase nmr studies, while of great importance, are beyond the scope of this work.

For high resolution nmr spectra a liquid sample can be examined as a pure liquid, or preferably, as a solution. Pure liquids may be viscous and high viscosity usually leads to line broadening (and so lower resolution) through restricted molecular motion (see Section 1.8). Similarly, solids must be dissolved in a suitable solvent to eliminate line broadening.

Π Can you think of some important properties that an nmr solvent might have?

For proton nmr spectra solvents must be *aprotic*, that is, the solvent must be free of protons itself. Thus deuterated solvents (Fig. 2.1a) are used or naturally aprotic materials such as carbon tetrachloride. Most commercial deuterated solvents have at least 98% isotopic purity but care has to be taken to recognise residual protonated solvent peaks in nmr spectra. Fig. 2.1a lists typical nmr solvents covering a range of polarities. Chloroform-$d_1$ is the most versatile and commonly used solvent.

When choosing a solvent you must be aware of potential solvent-solute interactions which might influence the nmr spectrum. Two important factors here, both of which can be concentration dependant, are *hydrogen-bonding*, and *proton exchange*. The chemical shifts of hydrogen-bonded protons (eg, the O—H or N—H protons in alcohols and amines) can vary enormously (4–5 ppm) depending on the concentration, and hence on the degree of hydrogen bonding

| Name | Formula | $\delta_H$* | Multiplicity** | $\delta_C$ | Multiplicity** |
|---|---|---|---|---|---|
| Carbon tetrachloride | $CCl_4$ | – | – | 96.0 | 1 |
| Carbon Disulphide | $CS_2$ | – | – | 192.8 | 1 |
| Chloroform-$d_1$ | $CDCl_3$ | 7.28 | 1 | 77.0 | 3 |
| Acetone-$d_6$ | $CD_3COCD_3$ | 2.07 | 5 | 29.8 | 7 |
| Dimethylsulphoxide-$d_6$ | $CD_3SOCD_3$ | 2.50 | 5 | 39.5 | 7 |
| Methanol-$d_4$ | $CD_3OD$ | 3.34 (4.11) | 5 | 49.0 | 7 |
| Pyridine-$d_5$ | $C_5D_5N$ | 7.2–8.6 | complex | 123.5 | 3 |
| Benzene-$d_6$ | $C_6D_6$ | 7.24 | 1 | 128.0 | 3 |
| Toluene-$d_8$ | $C_6D_5CD_3$ | 2.3, 7.1 | complex, 5 | 21.3, 125–137 | 7, complex |
| Acetic Acid-$d_4$ | $CD_3CO_2D$ | 2.06 (12) | 5 | 20.0, 178.4 | 7,1 |
| Trifluoroacetic Acid | $CF_3CO_2H$ | (12) | 1 | 115.0, 163.0 | 1,1 |
| Deuterium Oxide | $D_2O$ | (4.61) | 1 | – | – |

* Chemical shift of residual proton signals. Those in brackets may vary significantly depending on concentration and hydrogen bonding. Others may vary slightly.

** Refer to Sections 1.5–1.7 if you are not sure about multiplicity.

**Fig. 2.1a.** *Commonly used aprotic nmr solvents*

in the solution. Also, chemically labile protons (again eg O—H and N—H protons) can exchange with one another. If the solvent contains a labile deuteron then, because of the usual excess of solvent, it can exchange with the sample's labile protons effectively removing them from the sample. This usually results in an enhanced signal from one of the solvent resonances.

In fact, this exchange phenomenon can be used to advantage in identifying labile protons. After obtaining a spectrum using $CDCl_3$ as solvent a drop of $D_2O$ can be added and the solution shaken. Labile protons exchange for deuterons and their signal disappears. A small peak for HOD (usually at about 4.6 ppm) appears. Fig. 2.1b shows a simple example of this effect.

The concentration of the sample should be sufficient to obtain an adequate signal-to-noise ratio (see Part 4). 1–10% weight to volume solutions are common. For proton work a tube with an outside diameter of 5 mm is used and a volume of about 0.5 $cm^3$ is required. Thus 5–50 mg of material is needed. Remember, though, that this is a rough guide. Modern FT spectrometers can analyse very much less sample (less than 1 mg). Also it is the concentration of the spin-active nucleus which is important (protons, carbon-13) not the overall concentration of material.

Π For example, comparing the two structures shown in Fig. 2.1c calculate the relative amounts of the two compounds required to give roughly equal concentrations of protons.

You should have found that about 8 times as much of the bromo compound is needed for the same overall signal intensity. Given the molecular masses as in Fig. 2.1c here is how you should have done the calculation:

$$\%H = \frac{4}{246} \times 100 = 1.62 \quad \text{for the bromo compound}$$

and $$\%H = \frac{4}{32} \times 100 = 12.5 \quad \text{for methanol}$$

$$\therefore \quad 12.5/1.62 = 8 \text{ times as much bromo compound needed.}$$

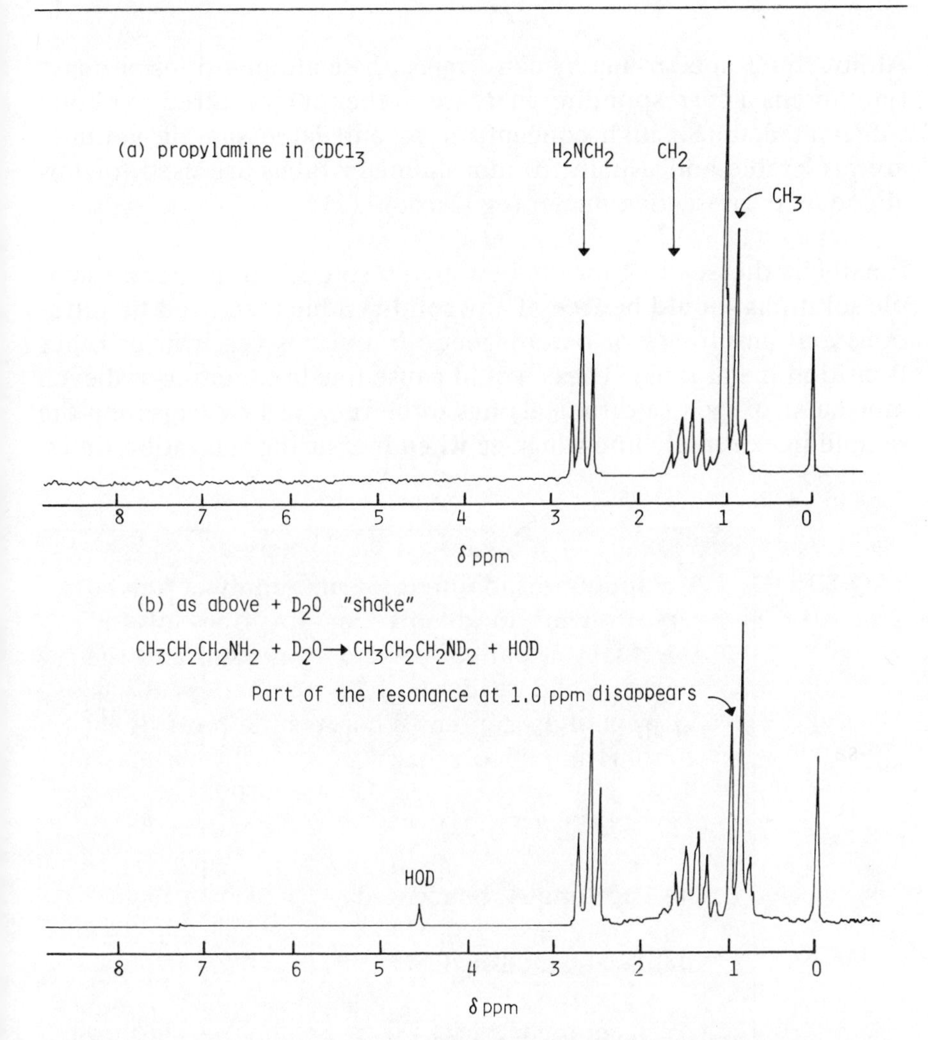


**Fig. 2.1b.** *Nmr spectrum on 1-aminopropane (n-propylamine)*

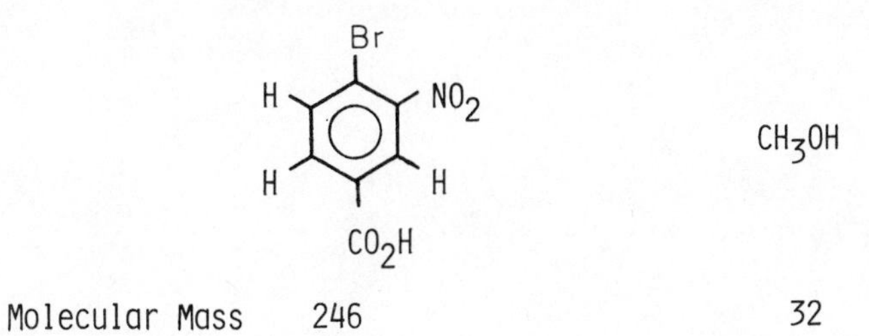


**Fig. 2.1c.** *Relative amounts of hydrogen in determining sample size*

Although FT spectrometers can cope with small quantities of material there is a corresponding increase in the time required to obtain sufficient data. So high concentrations and large sample volumes are preferable and usually 10 mm diameter tubes are used for low abundance spin-active nuclei (eg, carbon-13)

Finally, in this section, for the best high resolution nmr spectra sample solutions should be free of any solid residue (removed by filtration) and any traces of paramagnetic impurities (eg iron or other transition metal ions). These would cause line broadening and even molecular oxygen (a diradical) has to be removed by degassing the sample in extremely fine work, eg when measuring relaxation times.

**SAQ 2.1a**

A compound submitted for nmr analysis proved very difficult to dissolve in any one solvent. Eventually a pair of solvents was used and the spectrum had residual solvent resonances at $\delta$ = 2.1 ppm and 4.6 ppm. What was the most likely combination of solvents?

(*i*) $CDCl_3/CD_3OD$

(*ii*) acetone-$d_6$/benzene-$d_6$

(*iii*) acetic acid-$d_6$/$D_2O$

(*iv*) acetone-$d_6$/$D_2O$

**SAQ 2.1b**

Consider the two structures below (both compounds are used extensively in the polymer industry) and answer the following questions as either true (T) or false (F).

$CO_2H$

H H

H H

$CO_2H$

X

$HO_2C(CH_2)_6CO_2H$

Y

(*i*) the % hydrogen in X is 8.05.

T / F

(*ii*) Sample Y will have to be examined at higher concentration than sample X to give comparable signal intensities in their proton nmr spectra.

T / F

(*iii*) For comparable carbon-13 nmr spectra solutions of the same molarity should be used.

T / F

(*iv*) The presence of acidic protons could be shown by treating $CDCl_3$ solutions with $D_2O$.

T / F

**SAQ 2.1c**

Choose the correct word from the list below to answer the following questions:

(*i*) Solids and viscous liquids do not give high resolution nmr spectra because of what kind of effects?

(*ii*) What are compounds containing no hydrogen atoms called?

(*iii*) What are *two* concentration dependant solvent–solute interactions?

(*iv*) The presence of what kind of impurities will lower the resolution of an nmr spectrum?

(*v*) What technique can often be used to remove labile protons in a sample?

Choose from:

(*a*) Paramagnetic

(*b*) Proton exchange

(*c*) Chemical shift

(*d*) Aprotic

(*e*) Hydrogen-bonding

(*f*) Spin-active nucleus

(*g*) Line-broadening

(*h*) $D_2O$ exchange. ⟶

**SAQ 2.1c**

## 2.2. INSTRUMENTAL ASPECTS

All nmr spectrometers, whether continuous wave (CW) or Fourier Transform (FT) or even the recently developed nmr imaging instruments used in medicine, have certain features in common. In this section we will study some of these features, but leave to one side the complex electronics that link the components.

The common features are (*a*) a magnet, including a device for stabilising the magnetic field, (*b*) a sample probe which houses the sample and allows resonance to take place, (*c*) a source and detector of radiofrequency radiation, and (*d*) a data recorder/display system.

## 2.3. MAGNETS

There are currently three types of magnet used in nmr spectrometers: permanent magnets, electromagnets, and superconducting solenoid magnets.

Whichever type is used it must be capable of giving an intense, stable magnetic field with a homogeneity of the order of parts per billion (ie, $10^9$, the American billion). By homogeneity we mean that the magnetic field must be as uniform as possible in the region of the sample.

The idea of a magnetic field being composed of 'lines of force' allows us a pictorial view of homogeneity. In Fig. 2.3, part (*a*) shows the lines of force as homogeneous, passing uniformly from one pole of the magnet to the other. There are no distortions. Parts (*b*) and (*c*) show two types of inhomogeneity, the magnetic field curving, and a non-uniform field. Such distortions lead to loss of resolution. There are usually electronic devices, called shim coils, built into the pole faces of the magnets. Adjustment of small electric currents through these coils allows the level of homogeneity to be optimised.

Permanent magnets are the type usually associated with routine 60 and 90 MHz instruments. For proton resonance with a radiofrequency of 60 MHz a field strength of 1.4 Tesla (14,000 Gauss) is re-

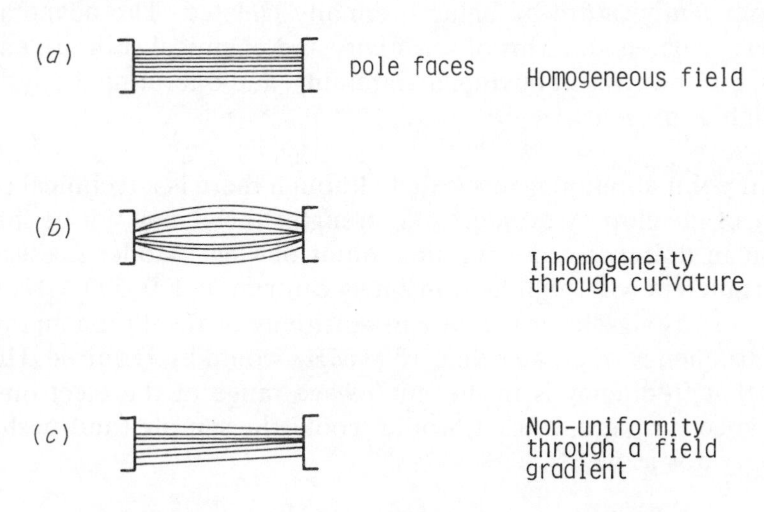


**Fig. 2.3.** *Homogeneity in magnetic fields*

quired while at 90 MHz 2.3 Tesla (23,000 Gauss) is needed. Modern permanent magnets have the advantages of providing stable magnetic fields and not requiring any power supply or cooling circuits, but they do require thermostating (as the magnetic field is temperature dependant) and cannot provide a large variable magnetic field. They can also be affected adversely by external magnetic fields and care has to be taken to shield or 'cage' such magnets.

Electromagnets are stable up to about 2.5 Tesla, and have the advantage of a variable magnetic field. However, they require a very stable power source to produce the high electric currents, and cooling circuitry must be supplied to keep the magnet at a constant temperature (± 0.1 °C). The cost of powering an electromagnet is considerable these days.

The recent development of superconducting solenoid magnets has extended the practical limit of stable, homogeneous magnetic fields up to about 12 Tesla, ie, nmr spectrometers working at about 500 MHz (for protons) can now be built. These are often referred to as high field nmr spectrometers and their magnets owe their remarkable strength to the superconducting properties of certain metals when cooled to extremely low temperatures (liquid helium, 3 K). The magnet is cryostatically controlled although the sample is held at room temperature by being thermally isolated. The advantages of these magnets in terms of sensitivity and chemical shift are enormous, but the costs of buying, maintaining, and operating the system are high at present.

A final point about magnets is that although there is a technical possiblity of developing even greater strengths there may not be much reason in doing so from the nmr point of view. Modern research spectrometers with high field magnets can run at 400–500 MHz but to obtain any signficant increase in sensitivity or resolution an operating frequency of greater than 1000 MHz would be required. However, that frequency is in the microwave range of the electromagnetic spectrum and, if used, would 'cook' the sample (and perhaps the operator as well!).

**SAQ 2.3a**

Which of the following best explains the term homogeneity as applied to magnetic fields?

(*i*) Homogeneity means how uniform the magnetic field is in the region of the sample.

(*ii*) Homogeneity means the magnetic field varies uniformly across the pole gap.

(*iii*) Homogeneity means the magnetic field is constantly varying as the nmr spectrum is being obtained.

(*iv*) A magnetic field is homogeneous when it is very stable.

**SAQ 2.3b**

Which of the statements below can be matched with:

(*i*) a permanent magnet

(*ii*) an electromagnet

(*iii*) a superconducting magnet?

*Note*: more than one statement may be appropriate for (*i*), (*ii*) or (*iii*).

(*a*) is used for high field nmr.

(*b*) is costly to run because of the price of electricity.

(*c*) is costly to run because of the price of liquid helium.

(*d*) provides stable magnetic field.

(*e*) is used for 60 MHz nmr spectrometers.

## 2.4. SAMPLE PROBES

The sample probe is the device which holds the sample, contained in its nmr tube, in between the poles of the magnet, in the strongest and most homogeneous part of the magnetic field. But as well as acting as a sample holder the probe has some other important functions. (Fig. 2.4a). Firstly, there is a *transmitter coil*, linked to the radiofrequency generator, which 'bathes' the sample in radiofrequency radiation of the appropriate frequency, say 90 MHz. (In FT spectrometers the radiation is 'pulsed', but the general idea is the same). A useful analogy here is of a radio station operating at 90 MHz (VHF) transmitting a radio programme.

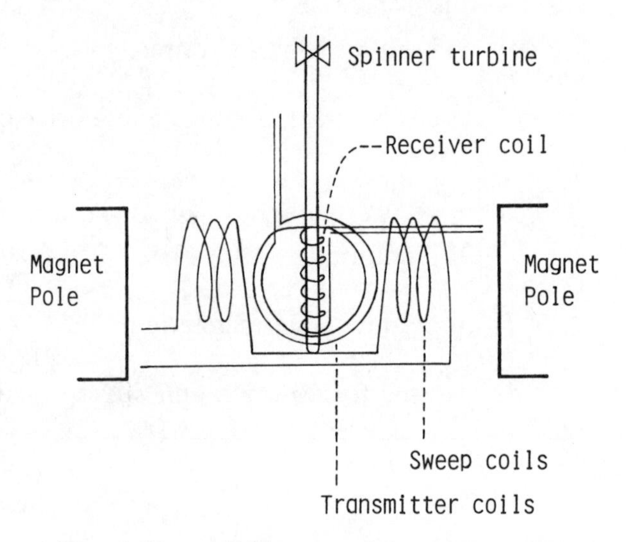


**Fig. 2.4a.** *Diagrammatic nmr probe*

Secondly, the probe has another coil, the *receiver coil*, wrapped in such a manner as to be at right angles to the transmitter coil. This is the optimum angle for detection of resonance for reasons we do not need to go into here. Our analogy continues with the receiver coil being likened to a transistor radio tuned to VHF, but having to be turned around to get the best reception.

Thirdly, in CW spectrometers there are a pair of *sweep coils*. (In some instruments these may not be part of the probe, but again it does not matter for our purposes). The sweep coils (known also as Helmholtz coils) are aligned so that a current through the coil generates a magnetic field *in the same direction* as the main magnetic field. Thus by changing the current in the sweep coils the magnetic field experienced by the sample can be varied or 'swept' so bringing all the spin-active nuclei into resonance. In our analogy the sweep coils would be the tuning device in a radio whereby we tune into a particular programme at a particular frequency, say 90 MHz.

In FT spectrometers there are no sweep coils. Instead the transmitter coils deliver high power *pulses* of radiofrequency radiation which excites all the nuclear spins at the same time. Section 4.3 discusses this in more detail.

Fourthly, the probe will allow the sample tube to be spun, usually by an air jet directed towards a spinner turbine. Spinning the sample along one axis effectively increases the homogeneity of the magnetic field over the sample, any differences being averaged out. Spinning the sample does lead to a problem however. So called *spinning side bands* appear symmetrically placed either side of the main spectral peaks and separated from them by the frequency of spinning (Fig. 2.4b). Usually these bands are easily recognised and reduced to low intensity (about 1% of the main peak) by careful maintenance and adjustment of the spectrometer. A low quality nmr tube might also cause spinning side bands, but usually such tubes do not spin at all.

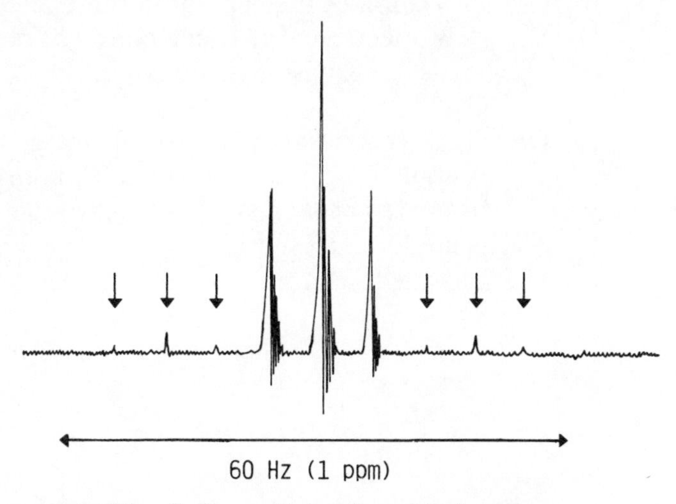


**Fig. 2.4b.** *Spinning side bands*

Measuring from this diagram the distance between the side bands and the main resonances is equivalent to 18 Hz. So the sample tube is spinning at this rate.

Fifthly, an nmr probe will usually have some means of regulating the temperature of the sample but not the magnet. (Not shown on Fig. 2.4b). These variable temperature probes (VTPs) allow the nmr experiment to be carried out over a range of temperatures (−100 ° to 200 °C) and so extend the versatility of nmr experiments to the study of kinetics, thermodynamics and so on.

Finally, there can be other features in a probe such as double resonance devices, but these are beyond the scope of this course.

**SAQ 2.4a**

For each of the four statements decide which of the choice of answers is most suitable:

(*i*) This allows nmr spectra to be obtained at high temperature: (*a*) heating the sample in a steam bath; (*b*) warming the magnet; (*c*) using a variable temperature probe.

(*ii*) Variation of the current in these leads to a CW spectrum: (*a*) sweep coils; (*b*) receiver coils; (*c*) transmitter coils.

(*iii*) This is a common cause of not getting a sample to spin in the probe: (*a*) faulty nmr tube; (*b*) broken sweep coils; (*c*) too much sample.

**SAQ 2.4b**

A single resonance in an nmr spectrum was found to have spinning side bands. If the resonance was at $\delta = 2.0$ ppm and the side bands at 2.3 and 1.7 ppm respectively and if the operating frequency of the instrument was 90 MHz, how fast was the tube spinning?

(*i*) 270 Hz

(*ii*) 90 Hz

(*iii*) 27 Hz

(*iv*) 54 Hz

## 2.5. RADIOFREQUENCY SOURCES AND DETECTORS

To appreciate radiofrequency sources and detectors in any depth would require detailed knowledge of electronics so our treatment will be highly descriptive rather than precise.

Radiofrequencies (Rf's) are generated by electronic multiplication of the natural frequency of a quartz crystal contained in a thermostated block. Different crystal sources and transmitters are used for different resonance frequencies. For example, a CW instrument operating at 60 MHz for protons would need a separate source of radiofrequency for fluorine-19 even though at the same magnetic field strength the resonance frequency of 56.4 MHz is similar to that for protons.

Radiofrequency sources are fitted with a power controlling device so that the level of irradiation can be varied. Up to a point, the greater the Rf power the greater the signal response, but you should recall from section 1.8 on relaxation effects that there is very little difference in population between spin energy levels. Hence, if we use a lot of power all the nuclei in the lower energy state are transferred to the high energy state and the nmr signals start to disappear. This leads in practice to distortion of the nmr signals—a phenomenon known as *saturation*. Complex equations relating optimum power level to sweep rate, sample volume, and many other factors can be derived, but in practice empirical adjustment of the Rf power is the easiest way of getting the best level.

For CW spectrometers the Rf power level is very small (about 1 watt), but with FT instruments where pulsing of the Rf occurs, power levels are much larger (typically 100 watts).

Detectors in nmr instruments have to be very sensitive as the signal levels are so small (less than 1 millivolt). Even so, multiple amplification of the signals is required before they are fed to the output devices. This amplification process is one source of electronic noise and contributes to the inherent low sensitivity of the nmr experiment. (Part 4).

**SAQ 2.5**

Which of the following statements about radiofrequency sources and detectors are correct?

Indicate your answer by circling T (true) or F (false).

(*i*) A typical Rf power level in CW nmr is 100 watts.

T / F

(*ii*) In nmr the term 'saturation' means that no more sample will dissolve in an nmr solvent.

T /F

(*iii*) Nmr detectors are not very sensitive.

T / F

## 2.6. DATA RECORDING AND DISPLAY

The sample in the nmr probe has yielded information or data which has been detected as an analogue signal, ie a voltage. This signal can now be fed to a number of output devices.

In routine analysis the signal is usually sent directly to a chart recorder to give a hard copy output or to an oscilloscope. The latter route is useful for fast scanning of the spectrum and allows preliminary adjustments to be made, eg setting optimum sensitivity or gain, and electronic filtering of the signal.

With modern nmr instruments, particularly FT spectrometers, the signal is sent to a computer via an analogue–digital interface. Within the computer the data can be processed in a number of ways, eg, to enhance signal-to-noise ratio, or, necessarily with FT instruments, to accumulate data and then mathematically transform it from a time-domain to frequency-domain spectrum. These points are discussed more fully in Part 4.

## 2.7. SUMMARY OF PART 2

Fig. 2.7a, a schematic diagram of an nmr spectrometer, summarises many of the points that have been discussed, and shows the relationship of the various components. The list below gives some key words and phrases associated with the main components of an nmr spectrometer.

Magnet – stable, homogeneous magnetic field provided by permanent, electro, or superconducting magnet.

Probe – houses sample, allows irradiation and detection of Rf radiation via transmitter and receiver coils. Also allows sweeping of magnetic field via sweep coils.

Rf source – quartz crystal sources, different ones required for different frequencies. Power level, saturation.

Data Recording and Display – chart recorder, oscilloscope and computer.

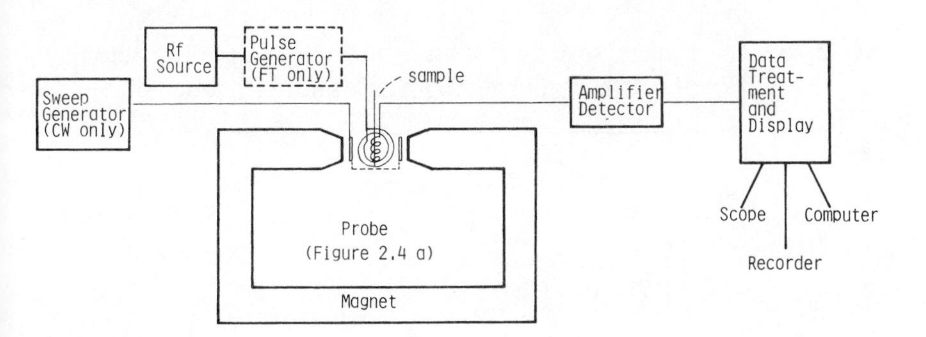


**Fig. 2.7.** *Schematic diagram of an nmr spectrometer*

**SAQ 2.7a**

Each of the following statements contains a word or phrase that is not necessarily correct. Pick that word or phrase out and substitute a correct word or phrase.

(*i*) After being detected, the nmr signal is automatically sent to a computer for processing.

(*ii*) Chart recorder output of an nmr spectrum occurs more quickly than an oscilloscope output.

(*iii*) Overall, an nmr spectrometer is quite a simple instrument.

**SAQ 2.7b**

Make your own sketch of the basic components of a CW nmr spectrometer and its probe.

**Objectives of Part 2**

Now that you have completed Part 2 you should be able to:

- describe aspects of sample preparation for nmr analyses;
- recognise that nmr solvents are aprotic, often deuterated materials;
- select suitable solvents taking account of solvent–solute interactions;
- explain the advantageous use of deuterium exchange of labile protons;
- assess the appropriate concentration required for a particular sample on the basis of its molecular mass and spin-active nucleus content;
- appreciate that low abundance spin-active nuclei like carbon-13 require different instrumentation from more abundant spin-active nuclei;
- describe the main instrumental features of an nmr spectrometer;
- list the main components of an nmr spectrometer as a magnet, a radiofrequency source, a probe, a detector system and a data processing system;
- state typical magnetic field strengths and corresponding radiofrequencies;
- explain the idea of homogeneity;
- compare the features of three different types of magnet in terms of field strength and sensitivity;
- appreciate that there is a practical limit to magnet size and strength;

- describe the main parts of an nmr probe as transmitter, receiver and, sweep coils (CW nmr), pulse generator (FT nmr)
- account for the phenomenon of spinning side bands;
- appreciate that an nmr spectrum can be obtained at different temperatures by using a suitable probe.
- appreciate that radiofrequency sources and detectors are complex electronic devices;
- recognise that CW spectrometers require low power radio frequency sources while FT instruments need a high power pulsed source;
- appreciate that nmr data can be processed in a number of ways;
- sketch the arrangement of the main components of an nmr spectrometer.

# 3. Qualitative Applications of NMR Spectroscopy

## Overview

Armed with some theory and the knowledge of how nmr spectra are obtained you can now try to extract and interpret the information contained in nmr spectra. We shall be concerned with qualitative information, relating nmr features to structural features, so you may want to revise the sections in Part 1 on chemical shift etc.

To help in these spectral interpretations we shall develop a set of rules-of-thumb, or heuristics (pronounced 'u-ris-tics'). First amongst these rules is the use of correlation charts and tables. You will be studying both proton and carbon-13 spectra so you will have to become familiar with correlation charts, etc, for both nuclei. Don't try to learn these things by heart, you would not be expected to do any serious interpretation without correlation charts. Nevertheless, by the end of this part you will probably have a good idea of where the proton and carbon-13 chemical shifts are for all the major functional groups.

Other heuristics you will encounter are to do with logical working, deducing structural facts from molecular formulae, and the effects of symmetry. In fact you should be able to list and explain at least six of these heuristics by the end of this part.

You should also be able to assign proton and carbon-13 spectra for given molecular structures and write reasoned and detailed interpretations of nmr spectra.

## 3.1. INTRODUCTION—CORRELATION OF SPECTRA

In this section we are going to study how the information contained in an nmr spectrum as chemical shifts, coupling constants and integration ratios can be converted to structural information. This *translation* of information or *interpretation* of spectral data involves the use of several logical and empirical 'rules-of-thumb' or *heuristics* (to use a modern, jargon word). We will see these ideas develop as we go through the examples, but there are two important points to grasp straight away.

Firstly, unlike, say, infrared spectroscopy all the peaks in an nmr spectrum can be assigned to structural features in the molecule under study. This statement may not be true for very complex molecules like proteins and nucleic acids, but as a heuristic it is very valuable because it tells us that when interpreting nmr spectra *we must try to account for every peak* that we see otherwise our analysis will be incomplete and our interpretation possibly incorrect.

Secondly, just as with translating languages, we need a 'dictionary' to translate nmr information into structural information. Like infrared and uv-visible spectroscopy such a 'dictionary' is provided by *correlation tables* or *charts*. In this work we shall use Fig. 3.1a and 3.1b to help interpret the proton chemical shifts and coupling constants. Fig. 3.1c gives roughly the same information in the form of charts for chemical shifts.

Fig. 3.1d provides a very useful correlation chart for carbon-13 chemical shifts.

| | | Methyl | Methylene | Methine |
|---|---|---|---|---|
| Substituents | | $-CH_3$ | $-CH_2-R$ | $-CH-(R)_2$ |
| Alkyl | HC— | 0.9 | 1.3 | 1.4 |
| Alkene | C=C— | 1.6 | 1.9 | 2.2 |
| Alkyne | C≡C— | 1.7 | 2.1 | 2.8 |
| Amide | $R_2N-CO-$ | 2.0 | 2.2 | 2.4 |
| Ester | RO—CO— | 2.1 | 2.2 | 2.5 |
| Ketone | R—CO— | 2.1 | 2.4 | 2.6 |
| Aldehyde | H—CO— | 2.2 | 2.3 | 2.4 |
| Cyano | N≡C— | 2.2 | 2.4 | 2.9 |
| Iodo | I— | 2.2 | 3.1 | 4.2 |
| Amine | $R_2N-$ | 2.2 | 2.5 | 2.9 |
| Phenyl | $C_6H_5-$ | 2.3 | 2.9 | 3.0 |
| Phenyl Ketone | $C_6H_5-CO-$ | 2.6 | 2.7 | 3.4 |
| Bromo | Br— | 2.7 | 3.3 | 3.6 |
| Amide | R—CO—NH— | 2.8 | 3.2 | 3.8 |
| Phenyl Amine | $C_6H_5-NH-$ | 2.9 | 3.1 | 3.6 |
| Chloro | Cl— | 3.0 | 3.6 | 4.0 |
| Ether | RO— | 3.3 | 3.4 | 3.6 |
| Alcohol | HO— | 3.4 | 3.5 | 3.9 |
| Ester | R—CO—O— | 3.7 | 4.2 | 5.1 |
| Phenyl Ester | $C_6H_5-CO-O-$ | 4.0 | 4.3 | 5.2 |
| Fluoro | F— | 4.3 | 4.4 | 4.8 |

**Fig. 3.1a.** *Proton chemical shifts for substituted alkanes* ($\delta_H \pm 0.3$ *ppm*)

| | 1 2 3 | $J_{11}$ | $J_{12}$ | $J_{13}$ | $J_{14}$ |
|---|---|---|---|---|---|
| Alkyl | >CH-CH-CH< | 12-15* | 6-8 | 0-1 | - |
| Alkenic | -CH=CH-CH< | 0-4 | 6-12 cis<br>12-18 trans | 0-3 | - |
| Alkynic | CH≡C—CH< | - | - | 2-3 | - |
| Aromatic | 1 2 3 4 (benzene ring) | - | 7-10 | 2-3 | 0-1 |

All $J$ values in Hertz

* Not observed for equivalent protons eg, in a $CH_3$- group.

**Fig. 3.1b.** *Typical proton–proton coupling constants*

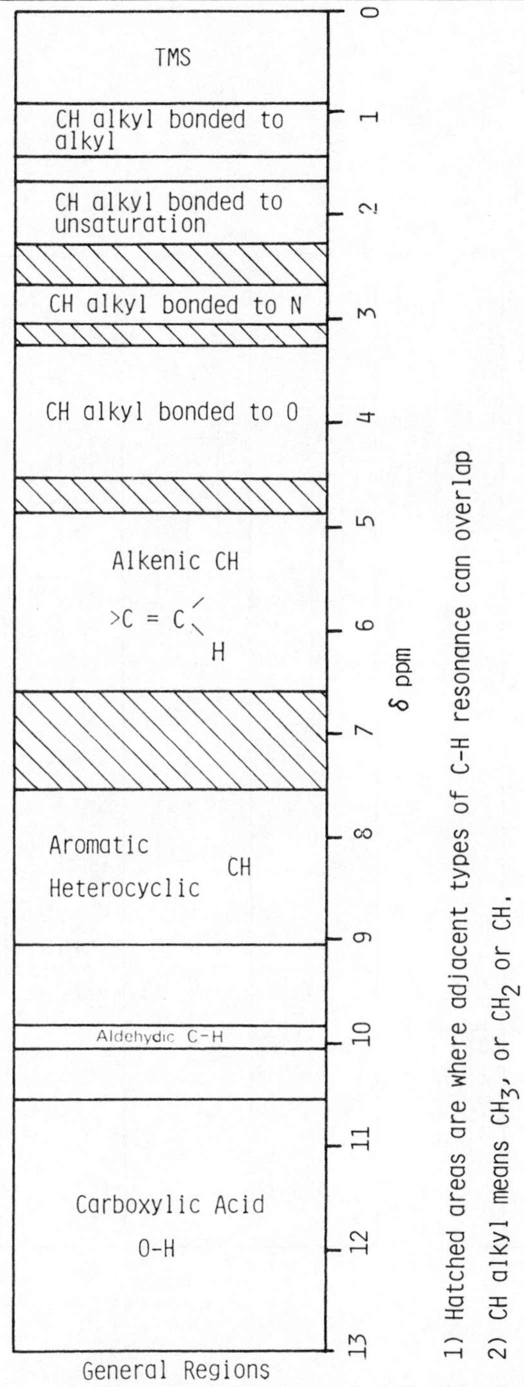


**Fig. 3.1c.** *A simple proton chemical shift correlation chart*

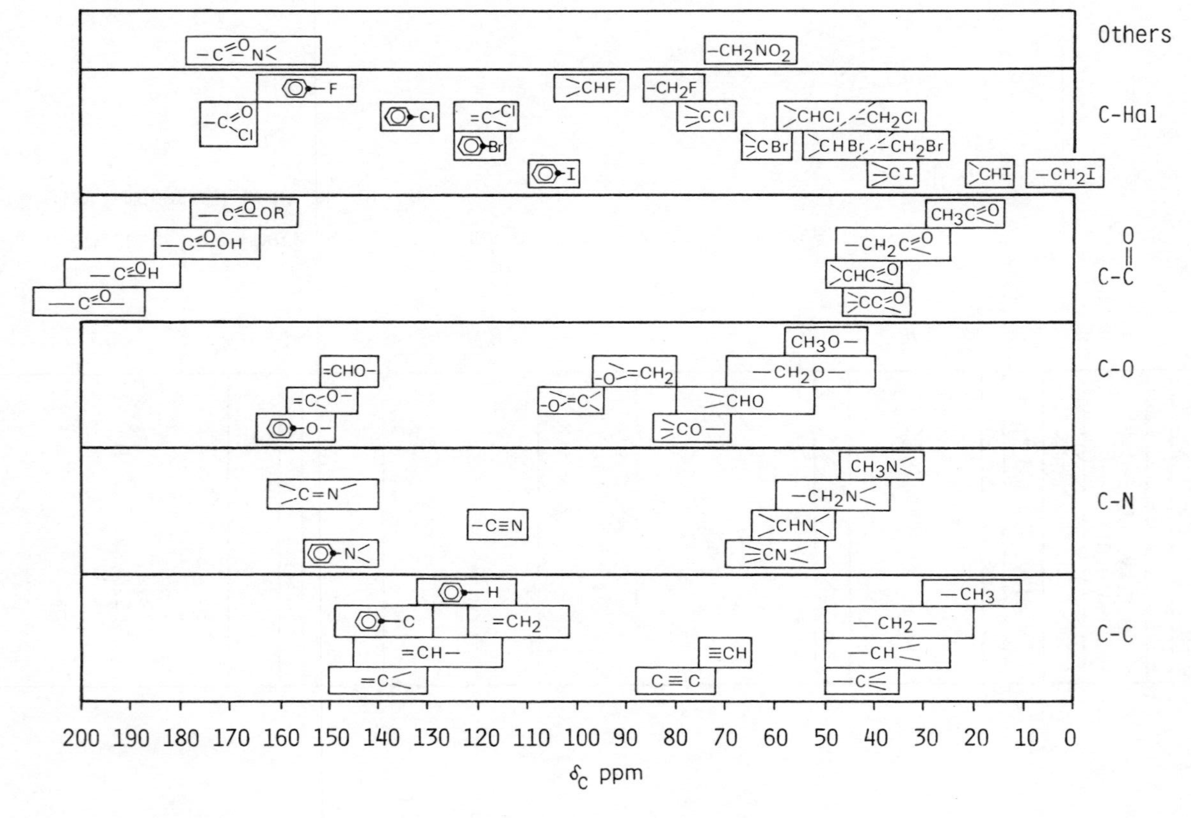


**Fig. 3.1d.** *Carbon-13 chemical shift correlation chart*

## 3.2. ASSIGNING A SPECTRUM—DIETHYL ETHER

We have already used the proton and carbon-13 spectra of diethyl ether to explore the meanings of the terms chemical shift, etc. Now we will employ them to learn how to use the correlation tables and charts and to *assign* a spectrum to a known structure. Fig. 3.2a shows these spectra.

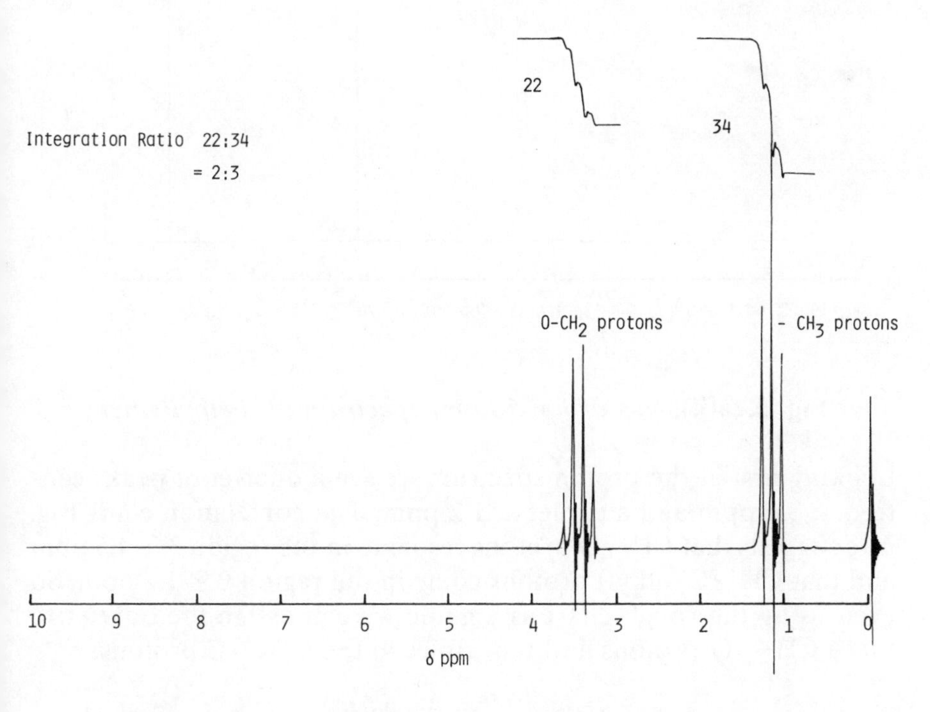


**Fig. 3.2a(i).** *Proton nmr spectrum of diethyl ether*

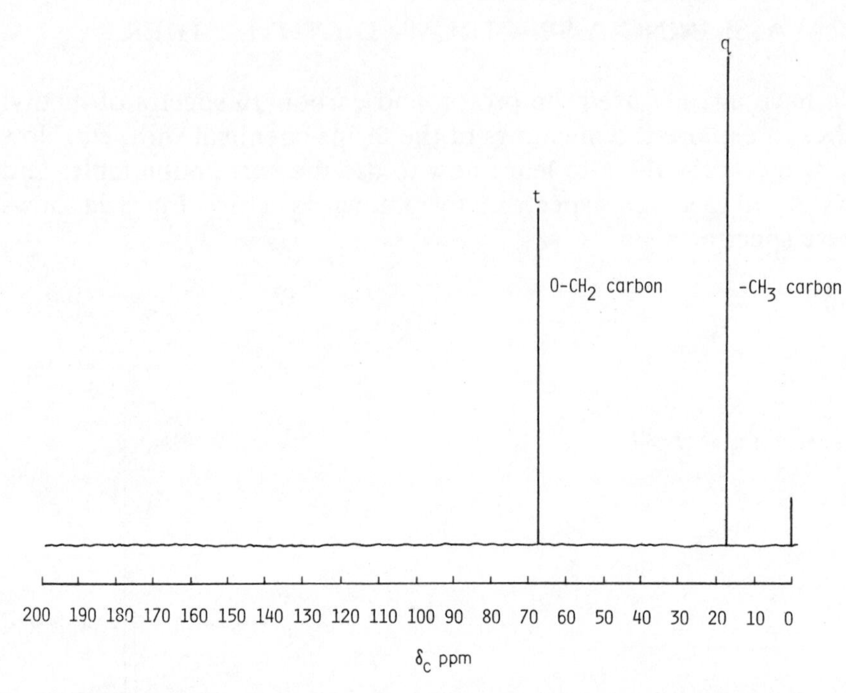


**Fig. 3.2a(ii).** *Carbon-13 nmr spectrum of diethyl ether*

Looking first at the proton spectrum we see a quartet of peaks centred at 3.4 ppm and a triplet at 1.2 ppm. The correlation chart Fig. 3.1c suggests that CH—O protons resonate in the region 3.2–4.5 ppm and that CH—C (alkyl) protons come in the region 0.9–1.2 ppm. So even using the rough chart as a guide we can assign the quartet to the $-CH_2-O$ protons and the triplet to the $CH_3-C$ protons.

Looking now at correlation table Fig. 3.1a we see an even closer match. The value 3.4 ppm is quoted for $CH_2-O$ protons. This is found by looking under the heading for a methylene group, ie $-CH_2-$alkyl. Then you should look down the corresponding *column* until you come to the value in the *row* corresponding to the ether substituent, RO—. Fig. 3.2b shows you this.

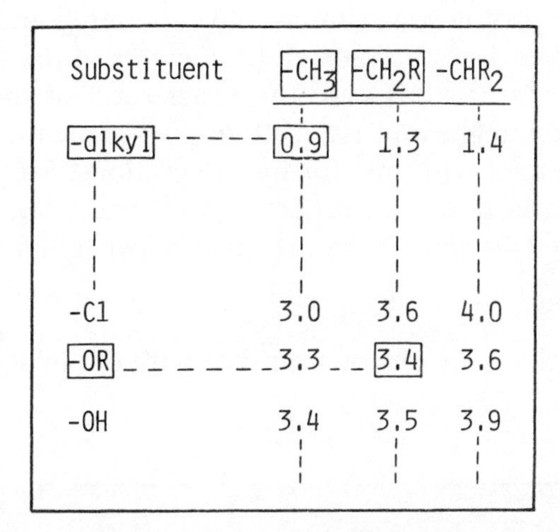


**Fig. 3.2b.** *Finding an entry in an nmr correlation table*

The same figure shows you how to get the expected value for the $CH_3$ protons in diethyl ether. The methyl group is bonded to a carbon and the value quoted is 0.9 ppm. However, there is an oxygen atom in the structure albeit one carbon removed. This has the long range effect of pulling the methyl group's resonance slightly downfield. Nevertheless, it is within the usually quoted range of $\pm$ 0.3 ppm. Not all correlations will be so exact.

Now let us consider the coupling constant and splitting pattern. We have already seen how to measure a coupling constant from a simple pattern such as this triplet and quartet. Check back to section 1.5 if you are not sure. In this case the value 6.5 Hz is found and consulting Fig. 3.1b for a typical value for an alkyl coupling between protons on adjacent carbon atoms, ie $J_{12}$, we find 6.5 Hz is well within the quoted range of 6–8 Hz. Fig. 3.2c explains some more terms used with coupling constants. So we have a good correlation of coupling constants with structure.

$J_{12}$ and $J_{23}$ are termed vicinal couplings

$J_{11}$, $J_{22}$, $J_{33}$ are termed geminal couplings

$J_{13}$ is a long range coupling, generally less than 1 Hz

**Fig. 3.2c.** *Coupling constant terms*

Finally, for the proton spectrum we have the integration ratio which is, as we found earlier, the ratio of the heights of the step curves on the nmr spectrum. In this case the exact ratio of the high to low field resonance integrations is 3.1 : 2, but we know that there can be an error associated with measuring integrations. So we can accept the observed ratio to be 3 : 2 reflecting $CH_3$ and $CH_2$. Note that this ratio cannot tell us that there are in fact two ethyl groups in the structure.

So we can match all the proton nmr data with the structure of diethyl ether.

**SAQ 3.2**

The carbon-13 spectrum for diethyl ether gives two peaks, at $\delta_C$ = 14.1 ppm and $\delta_C$ = 64.8 ppm. In the off-resonance spectrum the first peak splits into a quartet, while the second becomes a triplet. Match these data with the structure.

I have dealt with this example in detail for it exemplifies the two rules-of-thumb mentioned earlier—account for all the data, and use correlation tables—and the approach lays the basis for dealing with spectra from structures you do not know.

## 3.3. WORKED EXAMPLE 1

Fig. 3.3a and 3.3b show the proton and carbon-13 spectra for our first unknown compound. A glance at both spectra suggests that the structure is unlikely to be complex as there are few resonances and apparently no coupling in the proton spectrum. Spend a few seconds looking over these spectra.

Starting with the proton spectrum we see two resonances; one sharp band at $\delta$ = 2.3 ppm and a broader singlet at 7.3 ppm. From the correlation chart (Fig. 3.1a) these clearly belong to aklyl and aromatic protons respectively; so we know a lot about this structure straight away.

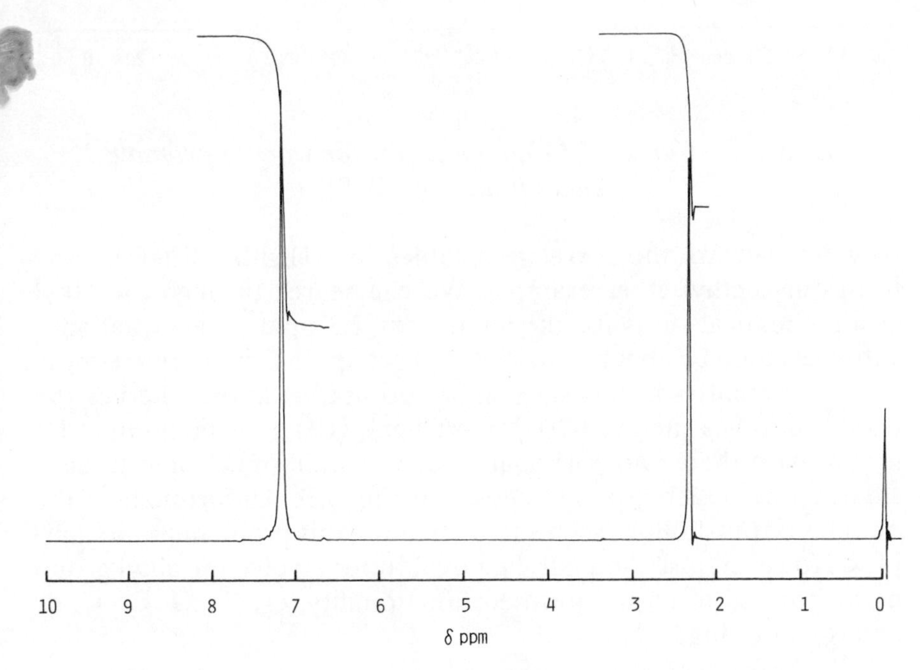


**Fig. 3.3a.** *Proton nmr spectrum for worked example 1 ($CDCl_3$ solution)*

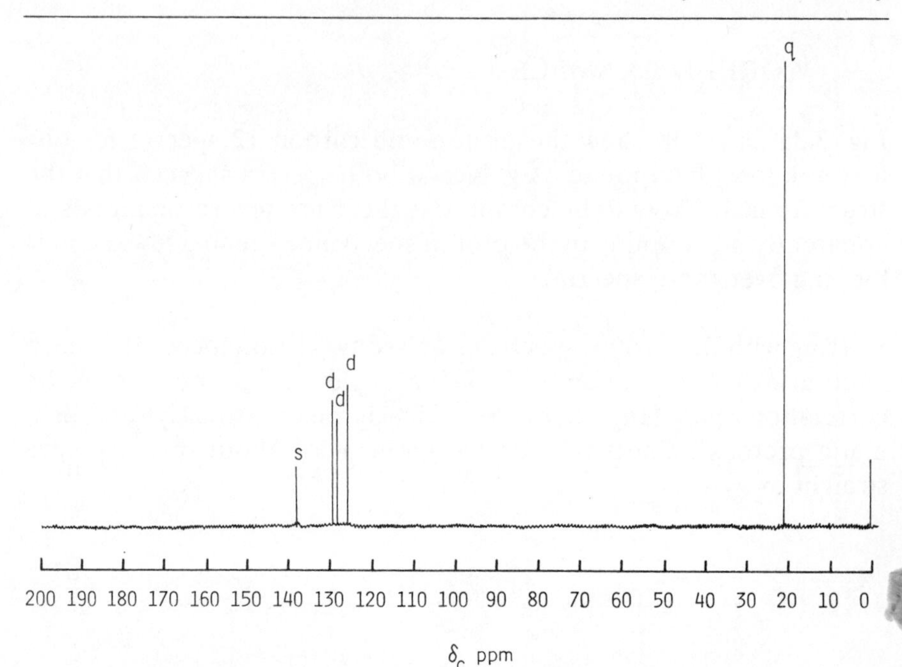


**Fig. 3.3b.** *Carbon-13 nmr spectrum for worked example 1 (neat liquid + TMS)*

Now we can use the correlation tables in a slightly different way from our diethyl ether example. We can search through the alkyl group chemical shifts for the value $\delta = 2.3$ ppm to see what substituents could be attached to the alkyl group. In fact, there are quite a few possibilities at this stage as we do not yet know whether the alkyl group is a methyl ($CH_3$), methylene ($CH_2$), or methine (CH) group. Also there is the normal error variation of 0.3 ppm to consider. Some possibilities are shown in Fig. 3.3c. Unfortunately, the number of possibilities seems to be making this problem more complex. However, just about all the possibilities involve the alkyl group being bonded to an unsaturated functionality, eg, C=O, C=C, or an aromatic ring.

To make progress let's look at the other information in the proton spectrum—the integration curves. Here the measured ratio seems to be 5 : 3 aromatic protons to aklyl protons. (Check this for yourself). Now this is fairly strong evidence that the aromatic portion of the structure is a mono-substituted benzene ring (that has five protons)

| $\delta$ ppm | | $\delta$ ppm | |
|---|---|---|---|
| 2.1 | $CH_3\text{-}CO_2R$ | 2.1 | $R\text{-}CH_2\text{-}C{\equiv}C\text{-}$ |
| 2.1 | $CH_3\text{-}COR$ | 2.3 | $R\text{-}CH_2\text{-}CHO$ |
| 2.3 | $CH_3\text{-}C_6H_5$ (benzene ring)  | 2.4 | $R\text{-}CH_2\text{-}C{\equiv}N$ |

**Fig. 3.3c.** *Some substituted alkyl protons resonating at $\delta = 2.3 \pm 0.3$ ppm*

and the alkyl portion is a methyl group (ie $-CH_3$). Note carefully that I said the evidence was 'fairly strong'. It is not conclusive as we are dealing with ratios and there might be ten aromatic and six alkyl protons. Nevertheless, returning to the chemical shift of the alkyl group and accepting the group to be a methyl we have narrowed substantially the number of possibilities. In fact considering the three structures in the first column of Fig. 3.3c we could say that the unknown is probably one of them, ie

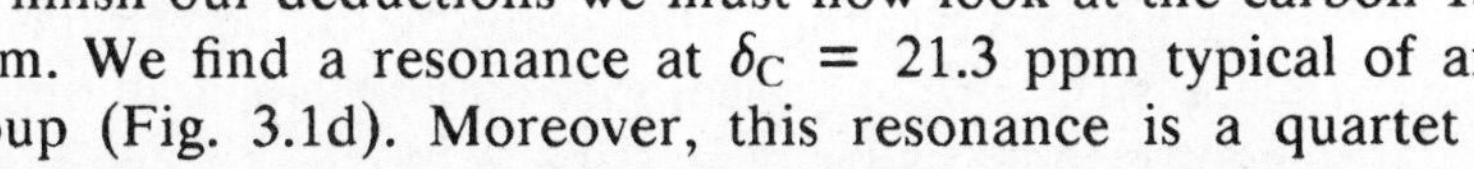

$CH_3\text{-}CO_2\text{-}C_6H_5$ — phenyl ethanoate

$CH_3CO\text{-}C_6H_5$ — acetophenone

$CH_3\text{-}C_6H_5$ — methylbenzene (toluene)

To finish our deductions we must now look at the carbon-13 spectrum. We find a resonance at $\delta_C$ = 21.3 ppm typical of an alkyl group (Fig. 3.1d). Moreover, this resonance is a quartet in the off-resonance spectrum, thus confirming the presence of a methyl group. There is a group of four peaks in the aromatic region (120–140 ppm) confirming the four different types of carbon atom in a mono substituted benzene ring (2 ortho, 2 meta, 1 para and 1 ipso carbon).

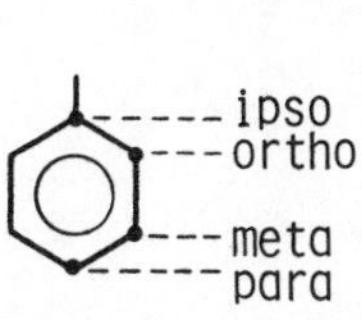

ipso
ortho
meta
para

Also, there is *no* peak corresponding to a carbonyl carbon atom (150–200 ppm). This in itself would rule out two of the above structures.

Thus summarising all our deductions the unknown structure was phenyl and alklyl groups with protons in the probable ratio of 5 : 3. This is substantiated by the carbon-13 spectrum which also tells us the aklyl is a methyl directly attached to the phenyl ring and that there is no carbonyl carbon.

So our first unknown compound is likely to be methylbenzene (toluene) (it is!). Note that we have not unambiguously assigned all the chemical shifts in the carbon-13 spectrum, but we have accounted consistently for all the data.

Now you're going to tackle a complete unknown for yourself. Approach this problem (and later ones) in the same manner that we used above in the first worked example. That is, look at the spectra; seeing what you can deduce from each spectrum in turn, using the correlation tables; write down what you can about one spectrum then the other, tabulating data if you think it helpful; and draw any conclusions. Don't worry if you don't get things absolutely correct first time, solving problems like this takes time and practice.

**SAQ 3.3**

Fig. 3.3d and 3.3e show the proton and carbon-13 nmr spectra for a pure compound. Determine the structure of the compound and assign all the resonances as far as possible.

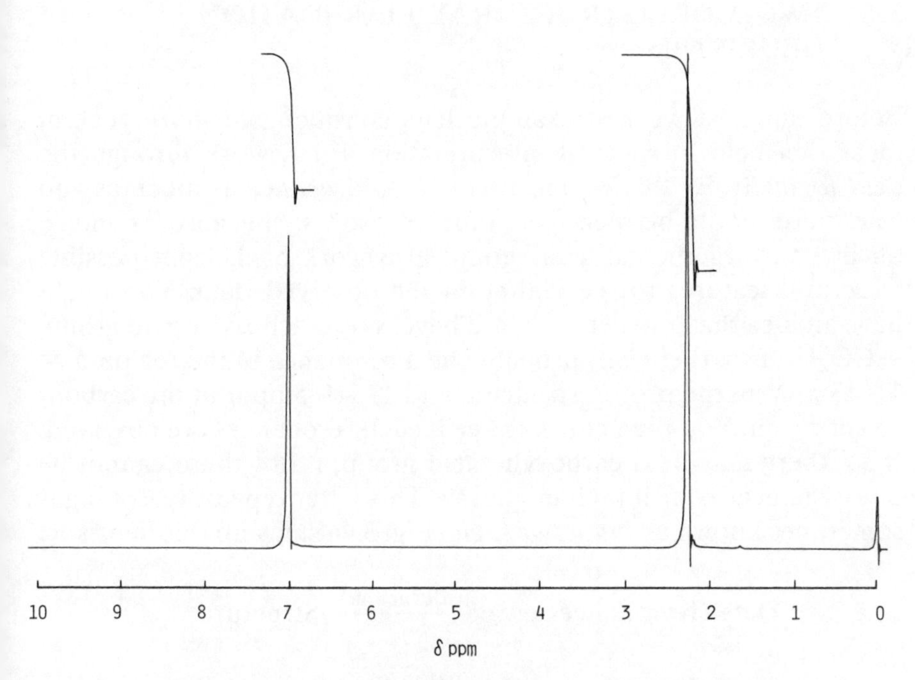


**Fig. 3.3d.** *Proton nmr spectrum for SAQ 3.3 ($CDCl_3$ solution)*

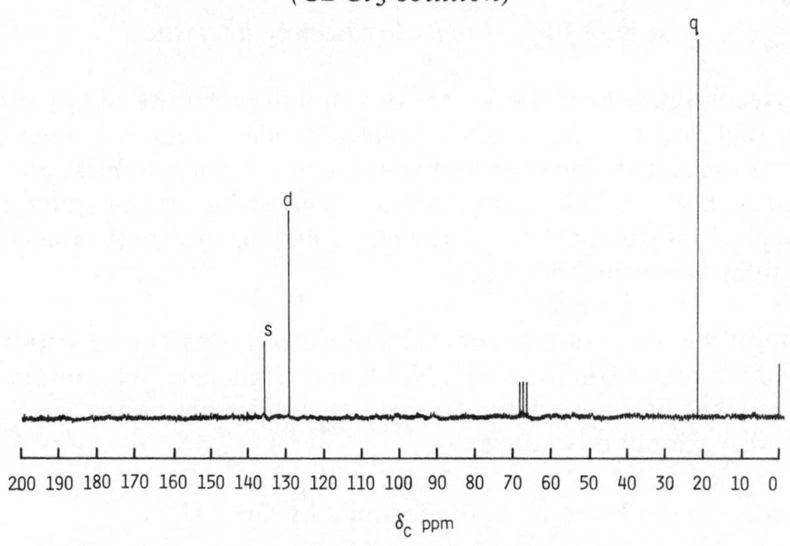


**Fig. 3.3e.** *Carbon-13 nmr spectrum for SAQ 3.3 ($CDCl_3$ solution)*

## 3.4. TWO MORE STRUCTURAL ELUCIDATION HEURISTICS

Before our next worked example let's consider two more general ideas that help in spectral interpretation. First, work through the data *logically*. By this we mean you should *deduce* as much as you can, there should be plenty of "thus" 's, "so" 's, "because" 's and so on in your arguments. You should also work back from possible structural features to see if they fit the observed data. You might have an idea that a structure could have, say a carboxylic acid group, $—CO_2H$. If so, there will probably be a resonance in the region $\delta$ = 11–13 ppm in the proton spectrum and 175–185 ppm in the carbon-13 spectrum. You then check to see if such resonances are observed. If so, there may be a carboxylic acid group, if not there cannot be one, whatever your hunch might say. This latter type of reasoning is sometimes known as *inductive logic*. Fig. 3.4a sums up this heuristic.

$$\text{Data (Resonances etc)} \xrightarrow{\text{Deduction}} \text{Structure}$$

$$\text{Possible Structures} \xrightarrow{\text{Induction}} \text{Check Data}$$

**Fig. 3.4a.** *Logical reasoning heuristic*

The second new heuristic is that you should make use of any other data that you can. Very often molecular mass data is known (or given in problems) and the molecular formula can also be supplied. Inspection of this latter information can give a surprising amount of structural information and a review of how to interpret molecular formulae is worthwhile here.

To limit the discussion to some extent we will consider only formulae which might contain C, H, N, O, and a halogen. We can obtain the number of double bond equivalents (ie double bonds + rings) by a very simple procedure:

Replace all the N atoms in the formula by CH

Replace all the O atoms in the formula by $CH_2$

Replace all the halogen atoms in the formula by H

Now using this new formula calculate the number of H's there would be in the fully saturated equivalent. The number of double bond equivalents is then half the calculated number of H's minus the observed number.

An example clarifies this:

$C_5H_4NO_2Cl \rightarrow C_5H_4(CH)(CH_2)_2(H)$ – replacement

$\rightarrow C_8H_{10}$

$\rightarrow C_8H_{18}$ – saturated equivalent of $C_8(2n + 2$ H's)

$\therefore$ 8H difference

$\therefore$ 4 double bond equivalents.

So this structure must possess the equivalent of four degrees of unsaturation, either as alkenic or carbonyl double bonds, rings, or triple bonds, or any combination.

**SAQ 3.4**

Choose for each of the molecular formulae below the correct number of double bond equivalents.

| | | |
|---|---|---|
| (*i*) | $C_4H_8$ | 1 or 2 or 4 |
| (*ii*) | $C_6H_6$ | 2 or 4 or 6 |
| (*iii*) | $C_5H_4O_2$ | 3 or 4 or 5 |
| (*iv*) | $C_5H_5N$ | 4 or 5 or 6 |

Let us see these two heuristics, logical reasoning and making use of a molecular formula, in action in the next worked example.

## 3.5. WORKED EXAMPLE 2

The proton and carbon-13 spectra for this compound are shown in Figs. 3.5a and 3.5b, and the molecular formula is $C_9H_{10}O_2$.

First, determine the number of double bond equivalents.

$C_9H_{10}O_2 = C_9H_{10}(CH_2)_2 = C_{11}H_{14}$

So $C_{11}$ would have 24 H's if fully saturated;

$\therefore$ 10 H's difference;

$\therefore$ 5 double bond equivalents.

(Note the logical deduction).

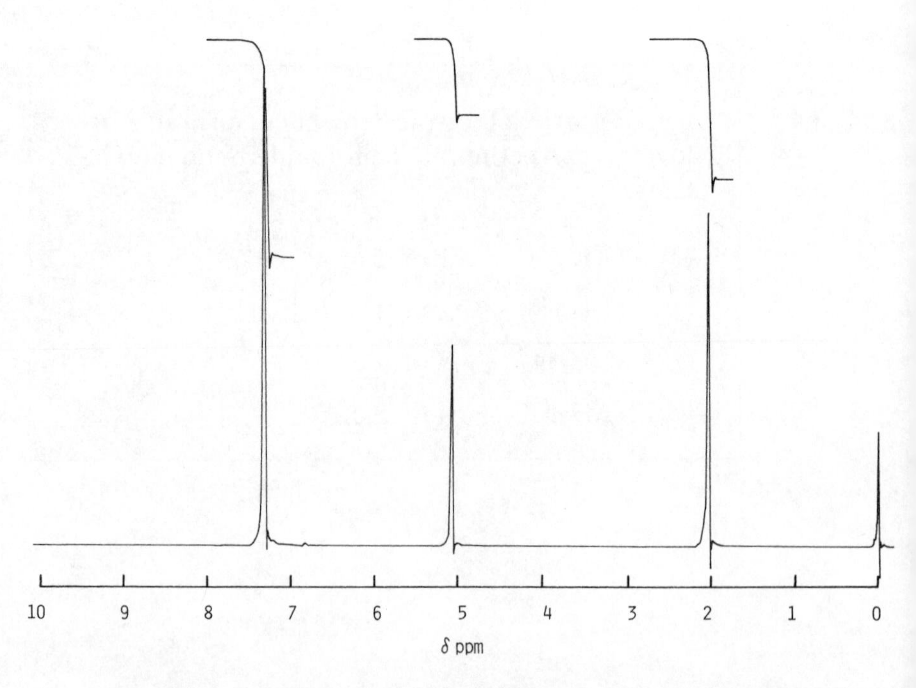


**Fig. 3.5a.** *Proton nmr spectrum for worked example 2 ($CDCl_3$ solution)*

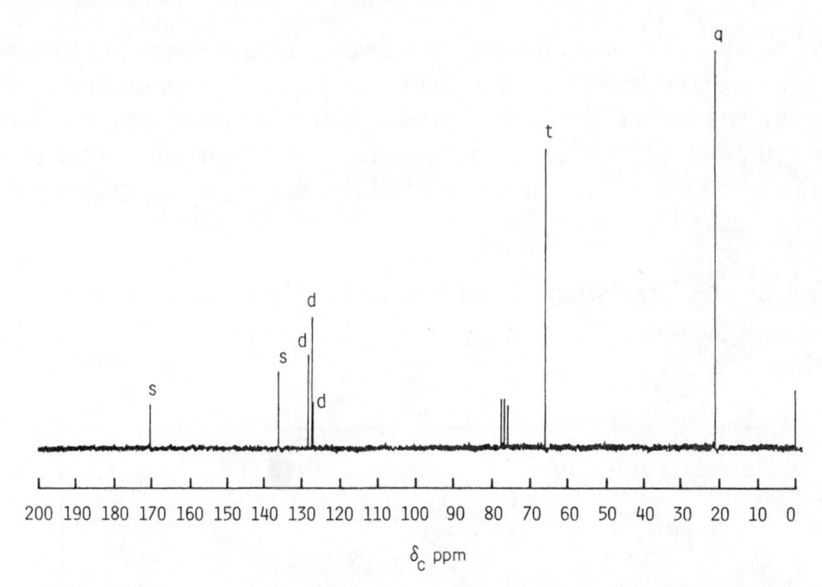


**Fig. 3.5b.** *Carbon-13 nmr spectrum for work example 2 ($CDCl_3$ solution)*

With such a large number of double bond equivalents and so few carbon atoms a possible structural feature could be a benzene ring. So we now check to see if this inference is correct by looking in the region 7–9 ppm in the proton nmr spectrum and 110–150 ppm in the carbon-13 spectrum. In both regions there are resonances so our inference is probably correct. (Strictly speaking there are other aromatic systems other than benzene rings which could resonate in those positions). That accounts for four of the five double bond equivalents.

Now consider the proton spectrum. There are three resonances at $\delta$ = 2.1, 5.1 and 7.3 ppm and no couplings. Correlation tables suggest an alkyl bonded to unsaturation for the resonance at 2.1 ppm and aromatic protons for the 7.3 ppm resonance. The resonance at 5.1 ppm falls in the alkenic region. Could this account for the remaining double bond equivalent? At this stage we cannot be sure. The integration ratio suggests protons in the ratio 3 : 2 : 5 (from high to low field), and this totals ten protons. So we have five aromatic protons—a monosubstituted benzene ring, and a methyl group. The two protons at 5.1 ppm are still not fully assigned.

The carbon-13 spectrum substantiates our deductions about the aromatic ring (see previous examples) and the methyl group and, using the correlation tables, it also shows us the presence of a carbonyl carbon $\delta_C = 170.5$ ppm as well as a methylene carbon at 66.1 ppm. Remember the off-resonance coupling data is presented on these spectra as letters (q – quartet, t – triplet, d – doublet).

We can now take stock of all our deductions and inferences, Fig. 3.5c.

| Structural Feature | Evidence δ ppm | |
|---|---|---|
| | Proton | Carbon-13 |
| 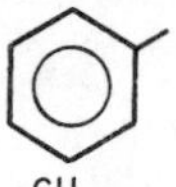 (benzene ring, monosubstituted) | 7.3 | 120-140 |
| $CH_3-$ | 2.1 | 20.7 |
| >C=O | - | 170.5 |
| $-CH_2-$ | 5.1 | 66.1 |

**Fig. 3.5c.** *Structural features in unknown 2*

The partial formula from the above features is $C_9H_{10}O$, so there must be another oxygen in the real structure as the molecular formula is $C_9H_{10}O_2$.

In fact we are in a position to decide the structure now, even though we could fit the molecular jigsaw together in several ways. Remember there is no proton coupling observed so each set of protons is isolated from the others. The methyl group must be attached to the carbonyl carbon (the chemical shifts demand this), the aromatic ring is also bonded to a carbon and this can only be the $CH_2$ group (again the chemical shift demands it) so the remaining oxygen *must* link the $CH_2$ to the carbonyl carbon.

Thus the structure is $C_6H_5CH_2OC(=O)CH_3$ benzyl ethanoate

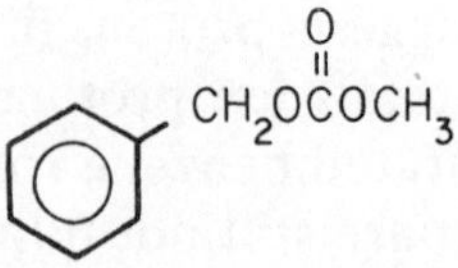

The methylene proton resonance at 5.1 ppm is explained by the group being bonded to two deshielding groups, oxygen and the aromatic ring.

Finally, we note that all the data is accounted for; there are no unexplained resonances in either spectrum.

**SAQ 3.5**

A compound of molecular formula $C_8H_8O$ has the following proton and carbon-13 spectral data. Deduce the structure of the compound.

Proton spectrum ($\delta$ ppm): 2.5 (3H); 7.4–8.0 (5H)

Carbon-13 spectrum ($\delta_C$ ppm): 26.3 (q); 128.2 (d); 128.4 (d); 132.9 (d); 137.1 (s); 197.6 (s)

The number of protons for each resonance is shown in brackets in the proton data as is the off-resonance splitting for the carbon-13 data, (s – singlet, d – doublet, q – quartet).

## 3.6 YET MORE STRUCTURAL ELUCIDATION HEURISTICS!

Don't worry, these next two rules-of-thumb are going to be the last that I shall discuss. The first new idea might have been apparent in the last example. While we tried to work logically *with* all the data we did not seem to work logically *through* the data, ie we jumped about from one spectrum and back again. Rather than being a fault this procedure has an advantage; we update our ideas as we go along, checking back to previous information to see if it still fits or adds more to our knowledge, then moving forward again. So the heuristic here is to constantly check and revise your reasoning.

The final heuristic concerns *symmetry* and *equivalence* in molecular structures. While not delving too deeply into what turns out to be a very tricky topic, you should look out for possible symmetry in molecular structures, and the general simplifying effect that symmetry can have on nmr spectra. The term equivalence is also used a lot in nmr spectroscopy and, for our purposes, we shall take it as referring to two or more groups whose chemical shifts are the same through an element of symmetry.

In fact we have already come across an example, diethyl ether, where the two ethyl groups are equivalent and the nmr spectra, both proton and carbon-13, cannot by themselves tell us that there are two equivalent ethyl groups.

Beware of the problem of symmetry in the next example.

## 3.7 WORKED EXAMPLE 3

Fig. 3.7a and 3.7b show the proton and carbon-13 spectra of unknown 3. Neither the molecular mass, nor the molecular formula is given. Can we obtain the structure of this compound?

Examine the proton spectrum.

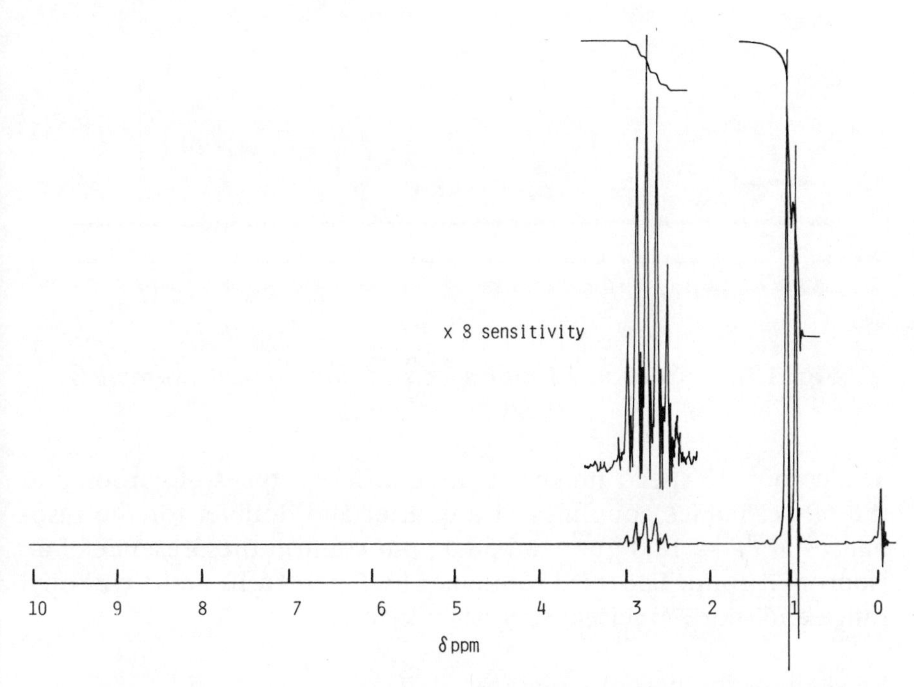


**Fig. 3.7a.** *Proton nmr spectrum for worked example 3 ($CDCl_3$ solution)*

Using the methods we have developed so far and taking especial note of the coupling pattern and integration ratio in the proton spectrum we can reason that there is an isopropyl group present in the structure (6 : 1 ratio, doublet and septet splittings, chemical shifts in the alkyl region). You might also suspect from the precise chemical shift of the methine proton that the isopropyl group is bonded to some unsaturation.

Now examine the carbon-13 spectrum.

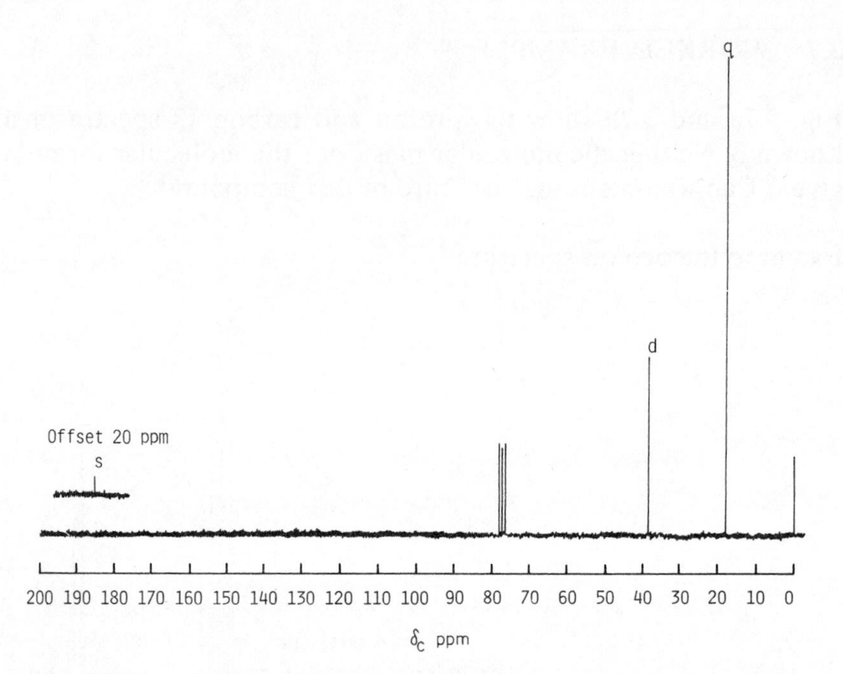


**Fig. 3.7b.** *Carbon-13 nmr spectrum for worked example 3 ($CDCl_3$ solution)*

The carbon-13 spectrum shows three different types of carbon and the off-resonance couplings of a quartet and doublet for the resonances at $\delta_C = 18.0$ ppm and 38.0 ppm confirm the presence of an isopropyl group. The third resonance (215 ppm) falls in the carbonyl range and more precisely suggests a ketone.

So we have the partial structure:

$$\begin{matrix} CH_3 \\ CH_3 \end{matrix} \!\!> CH{-}CO{-}R$$

What can R be if the structure is a ketone? The answer that springs to mind is another isopropyl group because, by symmetry, any other carbon containing group would give extra carbon-13 resonances, and probably proton ones as well.

After all of that can we be sure the compound is 2,4-dimethylpentan-3-one (diisopropyl ketone)? Not with the same certainty as other examples, although you may spend a long time trying to think of anything else that fits all the known data.

Π An ever hopeful student suggested the following structures that could perhaps fit the data of unknown 3. Why can they be discounted?

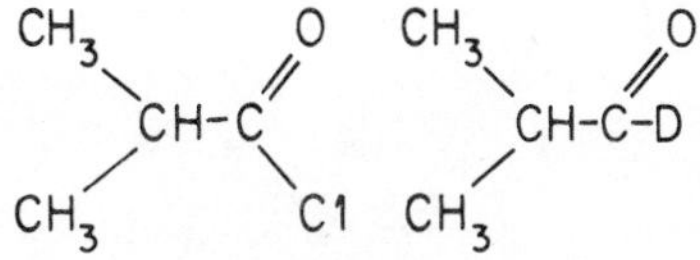

```
CH3       O    CH3         O
   \     //       \       //
    CH-C           CH-C-D
   /    \         /
CH3      Cl    CH3
```

For the acid chloride the carbonyl carbon chemical shift is wrong and for the deuteroaldehyde the deuterium would couple to the carbonyl carbon and give a 1 : 1 : 1 triplet for the resonance.

However, there is another possibility. Keep thinking and remember symmetry!

In the following question you might find it useful to start with the carbon-13 data. You don't always have to start with the proton spectrum. In fact we can propose another heuristic here. (I know I said we had finished with new heuristics in the last section, but, well, this one just seems to have crept in !). The absolutely final heuristic is: choose the simpler spectrum (data) to work on first. Remember though that this is a rule-of-thumb—it does not have to be obeyed.

**SAQ 3.7**

The proton and carbon-13 spectra of a compound of unknown structure containing only C, H and O are shown in Fig. 3.7c and 3.7d respectively. Deduce the structure of the compound.

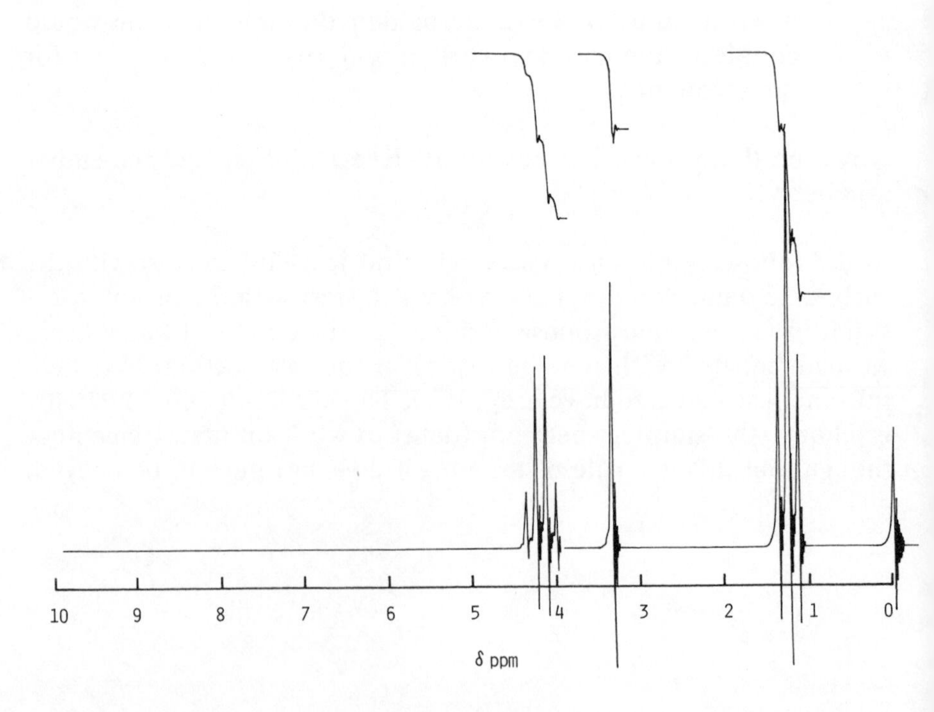


**Fig. 3.7c.** *Proton nmr spectrum for SAQ 3.7 ($CDCl_3$ solution)*

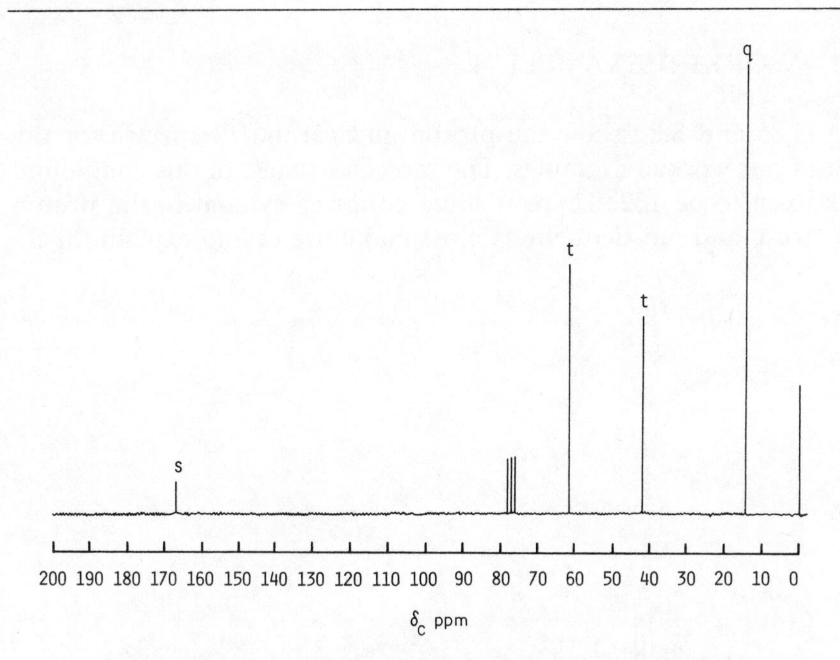


**Fig. 3.7d.** *Carbon-13 nmr spectrum for SAQ 3.7 ($CDCl_3$ solution)*

## 3.8 WORKED EXAMPLE 4

Fig. 3.8a and 3.8b show the proton and carbon-13 spectra for the last of our worked examples. The molecular mass of this compound is known to be 132. There is some coupling evident in the proton spectrum and our deductions must make use of and explain this.

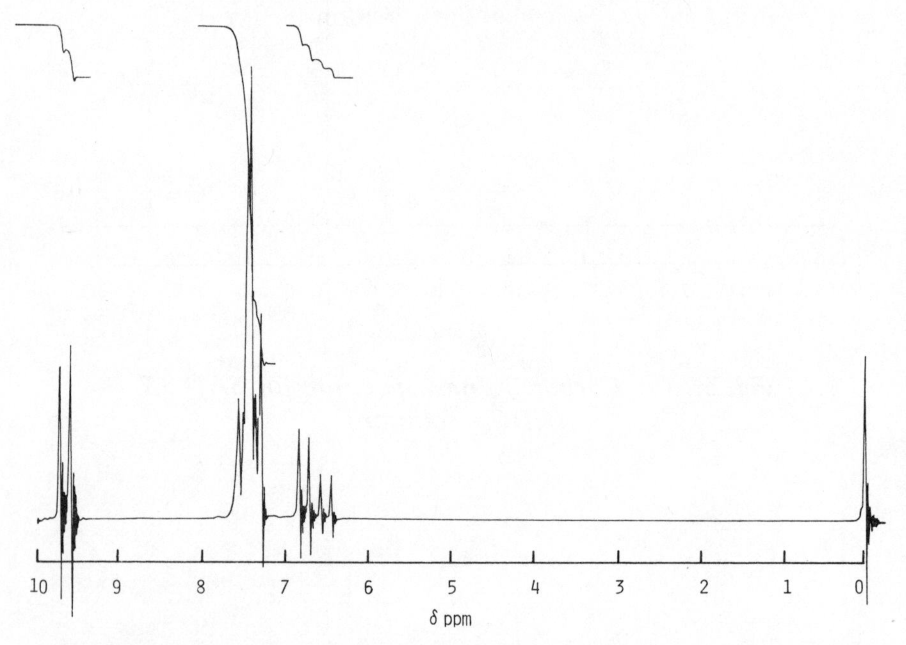


**Fig. 3.8a.** *Proton nmr spectrum for worked example 4 ($CDCl_3$ solution) molecular mass = 132*

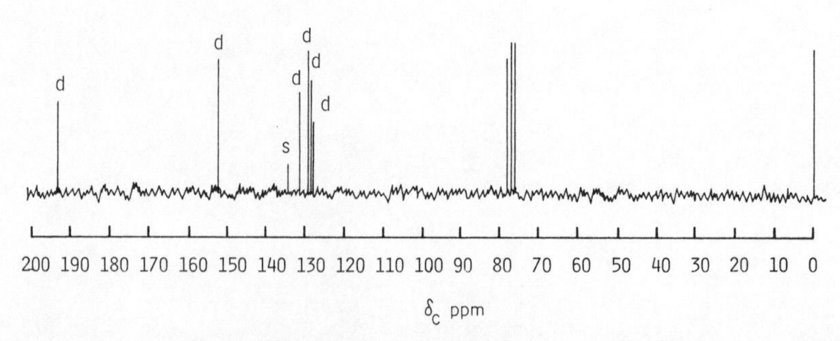


**Fig. 3.8b.** *Carbon-13 nmr spectrum for worked example 4 ($CDCl_3$ solution)*

Starting with the proton spectrum we can see three regions of resonance; the first doublet at $\delta$ = 9.7 ppm which the correlation chart (Fig. 3.1a) tells us must come from an aldehydic proton; the second a complex set of peaks around 7.4 ppm suggestive of aromatic protons; and the third a doublet of doublets centred about 6.7 ppm which suggest an alkenic proton. The ratio of intensities is 1:6:1 (make sure you check this for yourself) which indicates that this structure must have a multiple of eight protons in it.

The carbon-13 spectrum looks quite fearsome, but remember that the three peaks at $\delta_C$ = 77 ppm are from the solvent, $CDCl_3$. The presence of a carbonyl is confirmed by the resonance at 193 ppm and that it is an aldehydic carbonyl because the off-resonance datum tells us that the resonance is a doublet. Careful counting of the rest of the spectrum shows us that there are six other types of carbon atom, all of them either alkenic or aromatic.

Thus, summing up so far Fig. 3.8c shows us what functionality and evidence we have.

| Structural Feature | | | Evidence | |
|---|---|---|---|---|
| | | | $\delta$ ppm | |
| | Mass | | Proton | carbon-13 |
| —CHO | 29 | (definite) | 9.7(d) | 193(d) |
| $C_6H_5$ | 77 | (probably) | 7.4 | 128–150 |
| —CH= | 13 | (definite) | 6.7(dxd) | 128–150 |
| —Ch= | 13 | (necessary) | 7.4 | 128–150 |
| | 132 | | | |

**Fig. 3.8c.** *Functionality and evidence for unknown 4*

You should notice that the table includes a rough guide to how certain we are of the various structural pieces. You should also notice that a second alkenic CH has been included. This must be in the structure although the evidence is hidden under or in other resonances. It must be there to complement the one alkenic carbon we *know* to be there. ('You can't have one without the other'—that would make a good song!)

Finally the table shows that we have accounted for all the molecular mass. So our unknown must have the following bits:

$$C_6H_5- \quad , \quad -CH{=}CH- \quad , \quad -CH{=}O$$

There is only one way to fit these:

$$C_6H_5-CH{=}CH-C({=}O)H$$

3-phenylpropenal
(cinnamaldehyde)

To complete the exercise we need to account for the coupling pattern. This is quite easy if we look at the unsaturated aldehyde fragment:

$$-CH_A{=}CH_B-C({=}O)H_X$$

The aldehydic $H_X$ is split into a doublet by $H_B$. In turn $H_B$ is split by $H_X$, and by $H_A$. So that accounts for the doublet of doublets at $\delta = 6.7$ ppm. Unfortunately, the resonances of $H_A$ are hidden by the aromatic ones at 7.4 ppm.

If you thought that was tough just try the next question, but there are some easier ones to come in section 3.10.

**SAQ 3.8**

Fig. 3.8d and 3.8e show the proton and carbon-13 nmr spectra of a compound of unknown structure which contains only C, H and O. Deduce the structure of this compound and explain all the resonances.

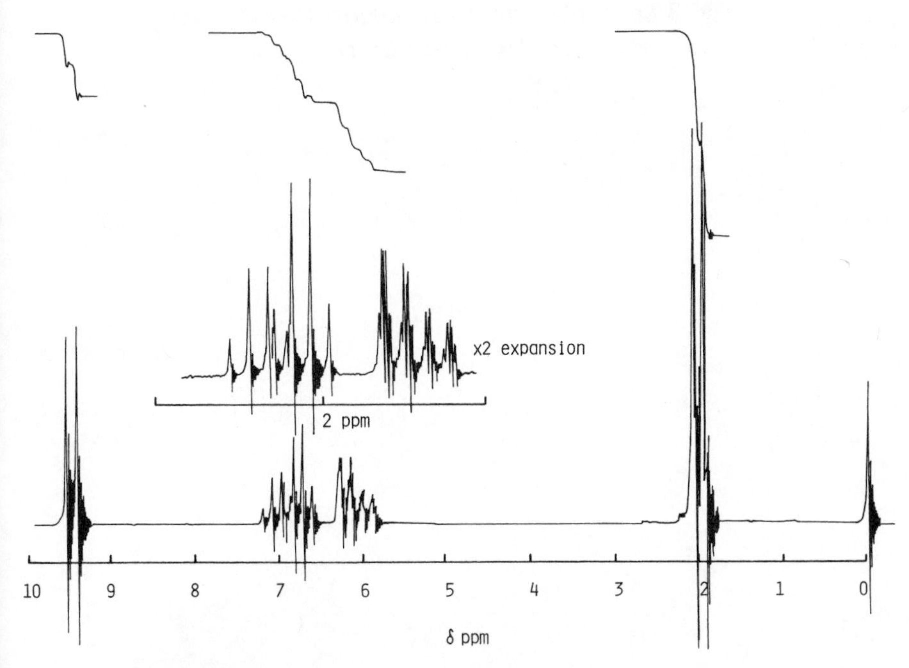


**Fig. 3.8d.** *Proton nmr spectrum for SAQ 3.8 ($CDCl_3$ solution)*

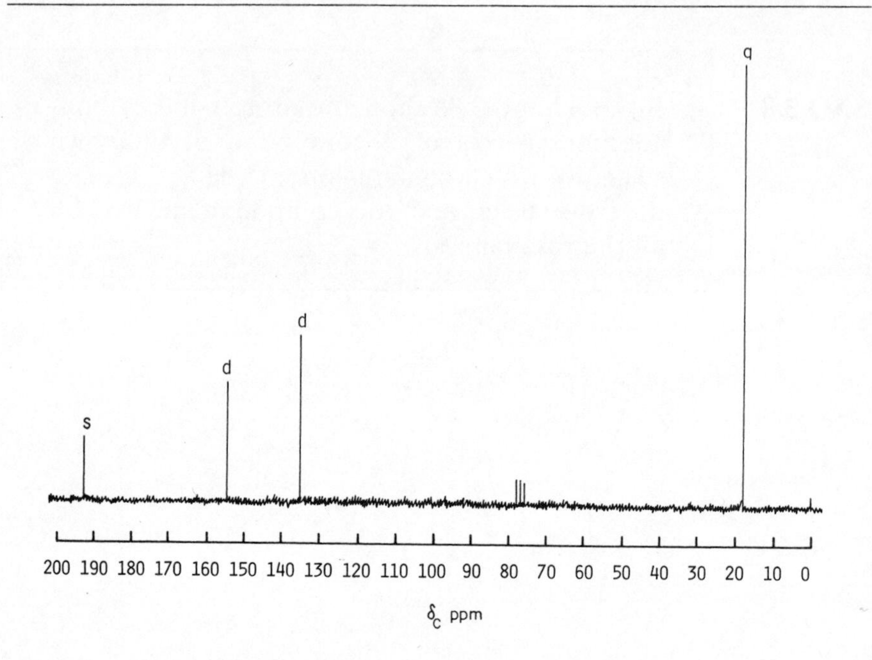


**Fig. 3.8e.** *Carbon-13 spectrum for SAQ 3.8 ($CDCl_3$ solution)*

## 3.9. SUMMARY OF PART 3

Before leaving you to attempt a set of problems it is worth summarising the main points, or heurisitics, that we developed in this part of the Unit.

(*i*) Try to account for all the data.

(*ii*) Use correlation tables/charts.

(*iii*) Work logically with the data.

(*iv*) Use any other data, particuarly a molecular formula, if given.

(*v*) Check your reasoning as you go along.

(*vi*) Beware, and make use of symmetry.

(*vii*) Choose the simpler spectrum for your first deductions.

**SAQ 3.9a to SAQ 3.9f: 6 problems**

Deduce the structure for each compound from the proton and carbon-13 spectra and other data that are provided. You should also explain your interpretation of the spectra as far as possible. The problems are loosely graded with the fairly easy ones first.

**SAQ 3.9a**

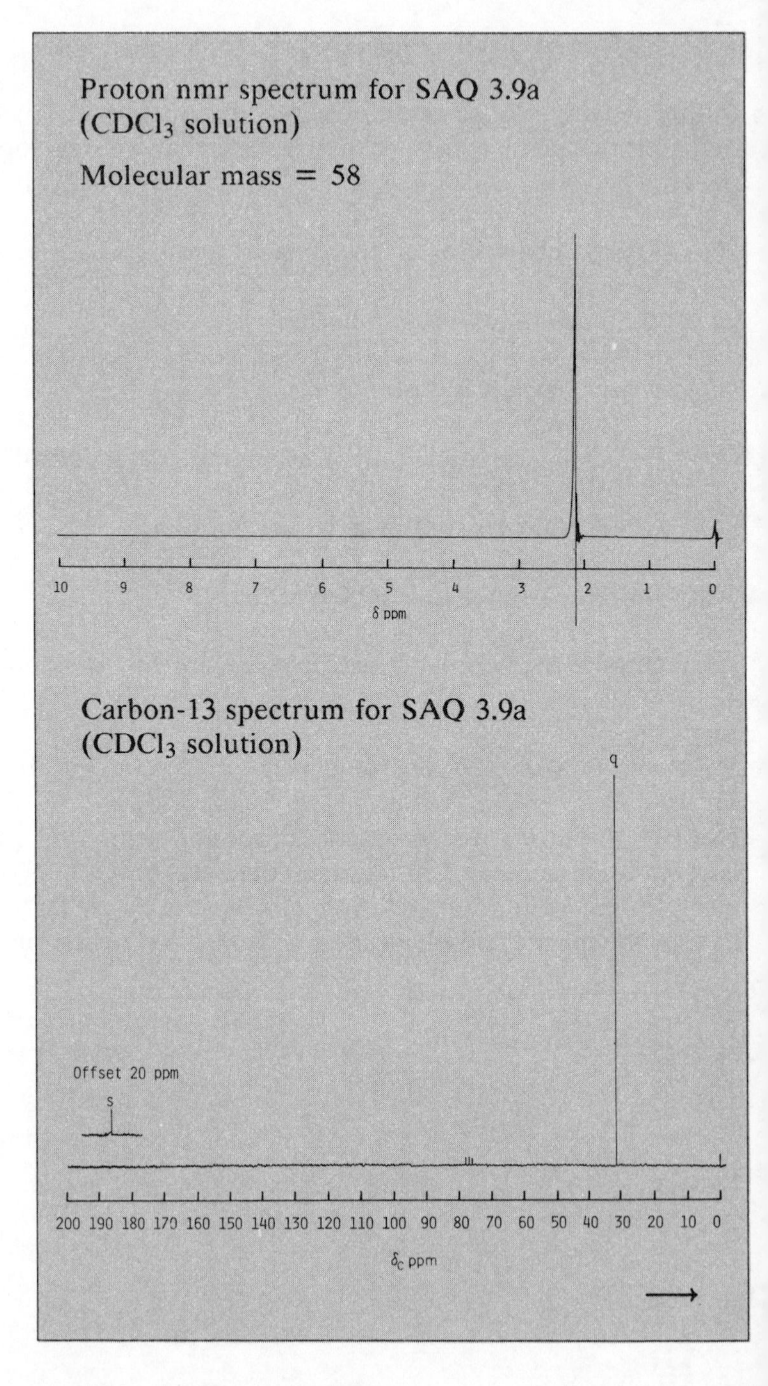
Proton nmr spectrum for SAQ 3.9a
(CDCl3 solution)
Molecular mass = 58
10 9 8 7 6 5 4 3 2 1 0
δ ppm
Carbon-13 spectrum for SAQ 3.9a
(CDCl3 solution)
q
Offset 20 ppm
s
200 190 180 170 160 150 140 130 120 110 100 90 80 70 60 50 40 30 20 10 0
δc ppm

**SAQ 3.9a**

**SAQ 3.9b**

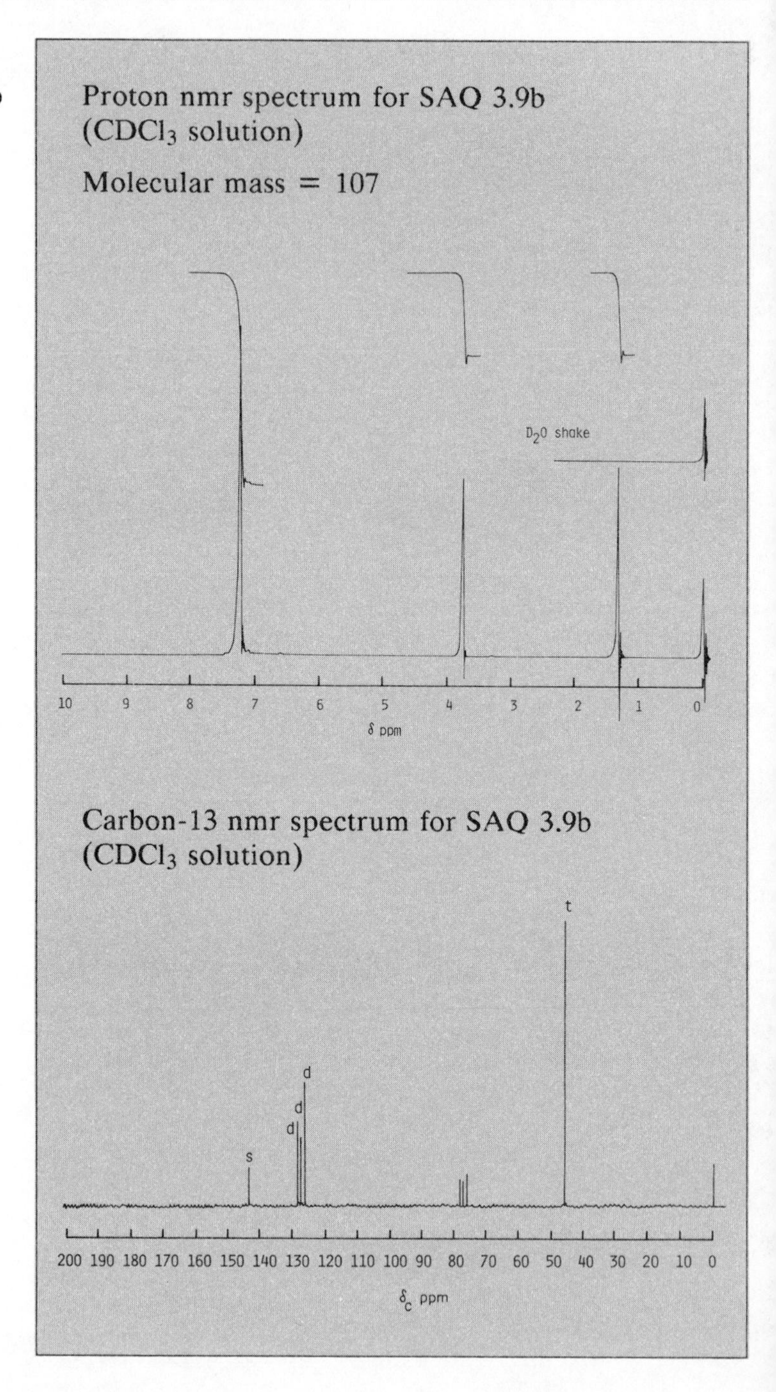
Proton nmr spectrum for SAQ 3.9b
($CDCl_3$ solution)
Molecular mass = 107
$D_2O$ shake
10 9 8 7 6 5 4 3 2 1 0
δ ppm
Carbon-13 nmr spectrum for SAQ 3.9b
($CDCl_3$ solution)
t
d
d
d
s
200 190 180 170 160 150 140 130 120 110 100 90 80 70 60 50 40 30 20 10 0
$\delta_C$ ppm

**SAQ 3.9b**

**SAQ 3.9c**

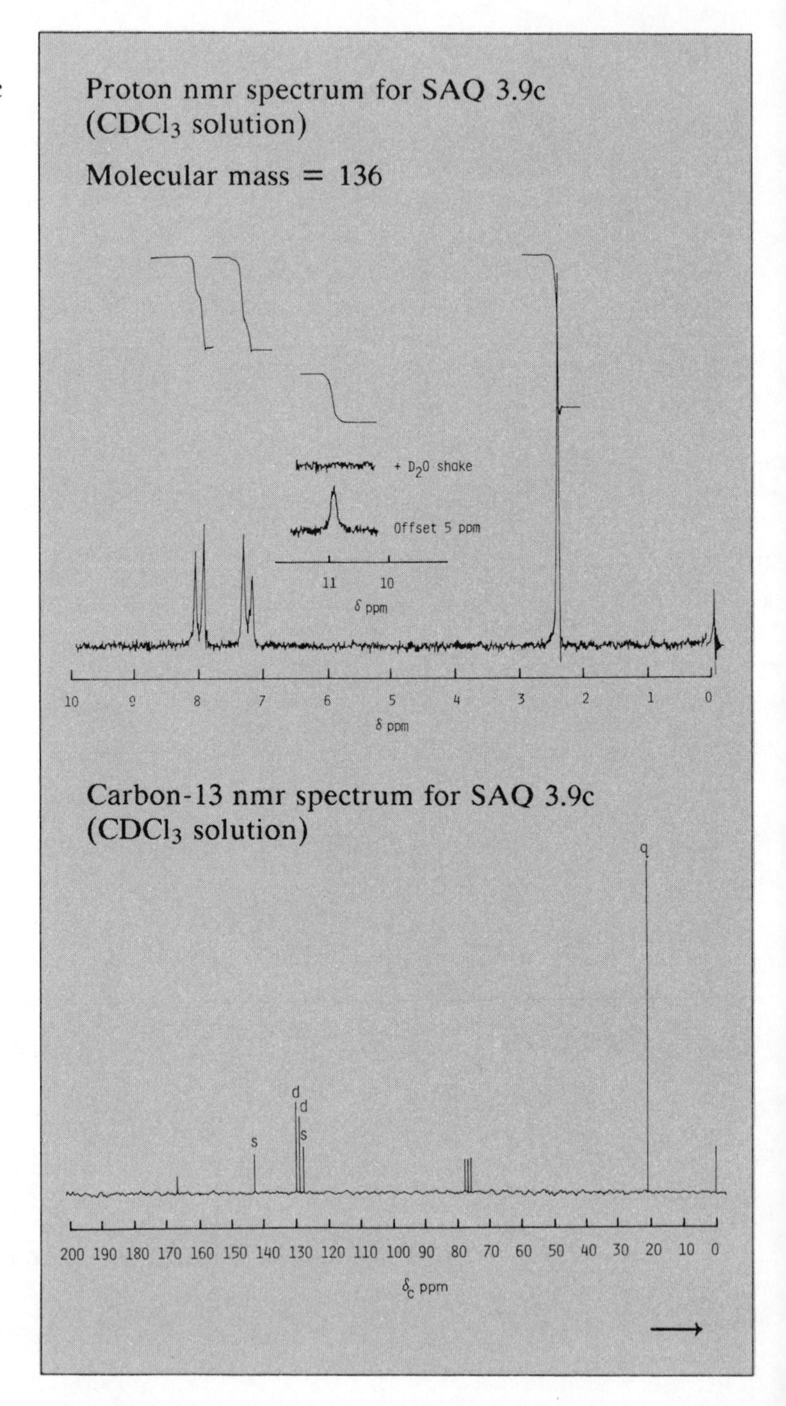
Proton nmr spectrum for SAQ 3.9c
(CDCl3 solution)
Molecular mass = 136
+ D2O shake
Offset 5 ppm
11
10
δ ppm
10
9
8
7
6
5
4
3
2
1
0
δ ppm
Carbon-13 nmr spectrum for SAQ 3.9c
(CDCl3 solution)
q
d
d
s
s
200 190 180 170 160 150 140 130 120 110 100 90 80 70 60 50 40 30 20 10 0
δC ppm

**SAQ 3.9c**

**SAQ 3.9d**

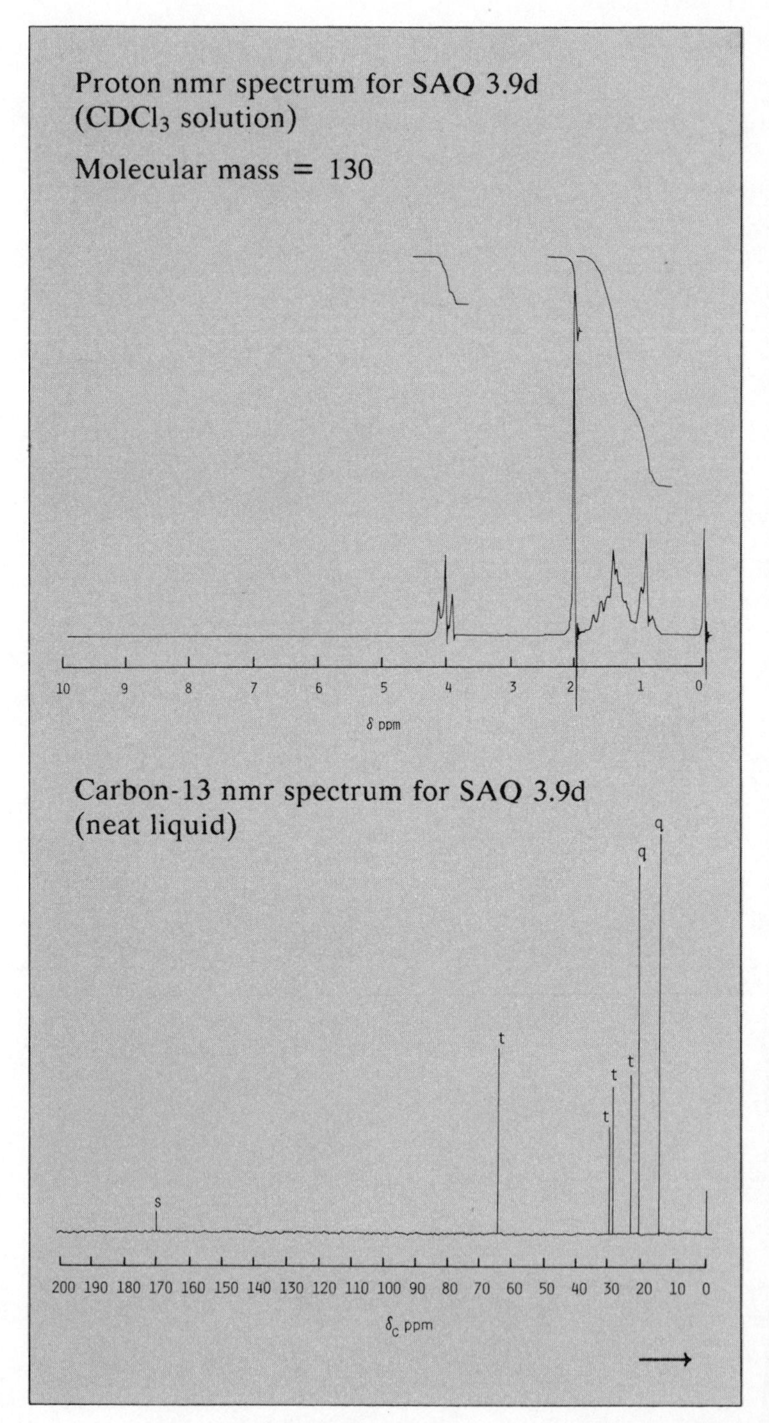
Proton nmr spectrum for SAQ 3.9d
(CDCl3 solution)
Molecular mass = 130
10
9
8
7
6
5
4
3
2
1
0
δ ppm
Carbon-13 nmr spectrum for SAQ 3.9d
(neat liquid)
q
q
t
t
t
t
s
200 190 180 170 160 150 140 130 120 110 100 90 80 70 60 50 40 30 20 10 0
δc ppm
⟶

**SAQ 3.9d**

**SAQ 3.9e**

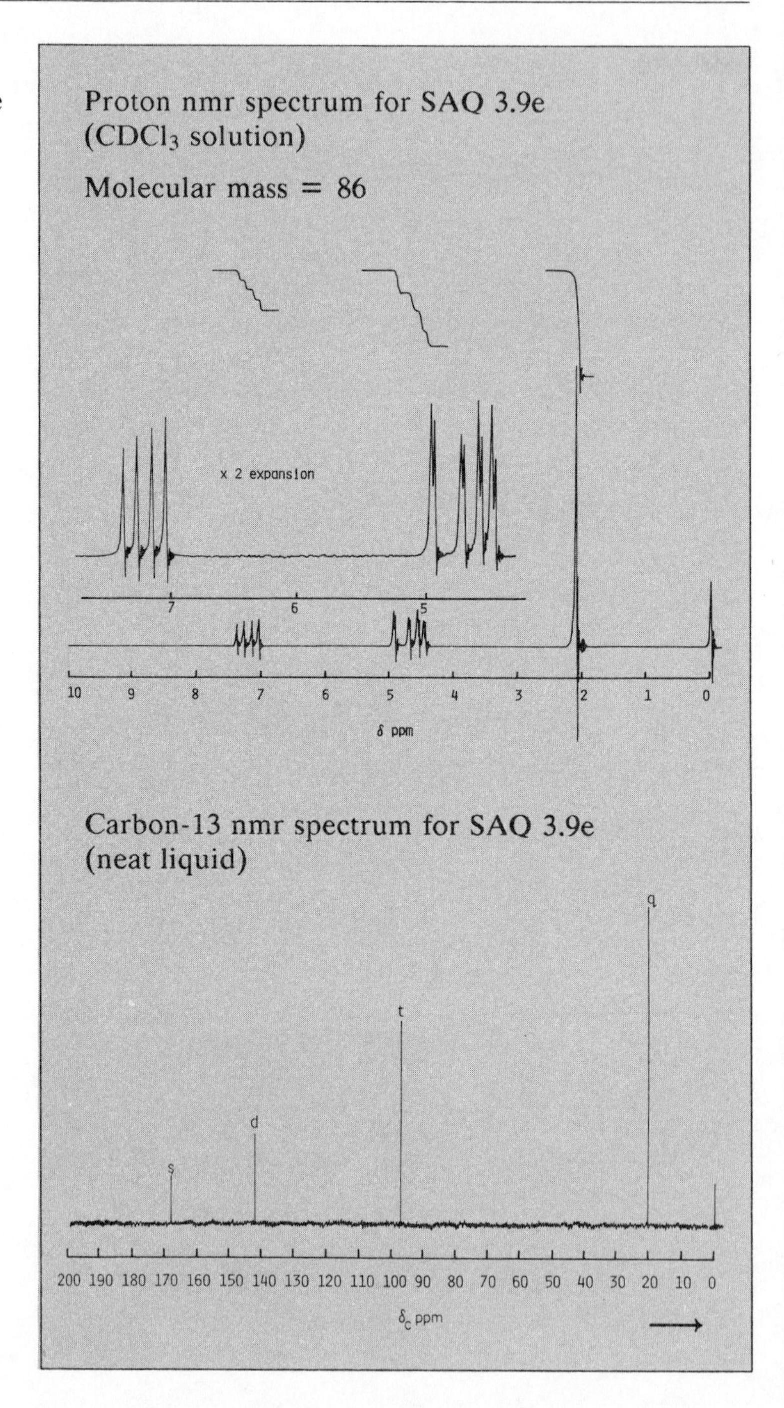
Proton nmr spectrum for SAQ 3.9e
($CDCl_3$ solution)
Molecular mass = 86
x 2 expansion
7
6
5
10
9
8
7
6
5
4
3
2
1
0
δ ppm
Carbon-13 nmr spectrum for SAQ 3.9e
(neat liquid)
q
t
d
s
200 190 180 170 160 150 140 130 120 110 100 90 80 70 60 50 40 30 20 10 0
$\delta_C$ ppm

**SAQ 3.9e**

**SAQ 3.9f**

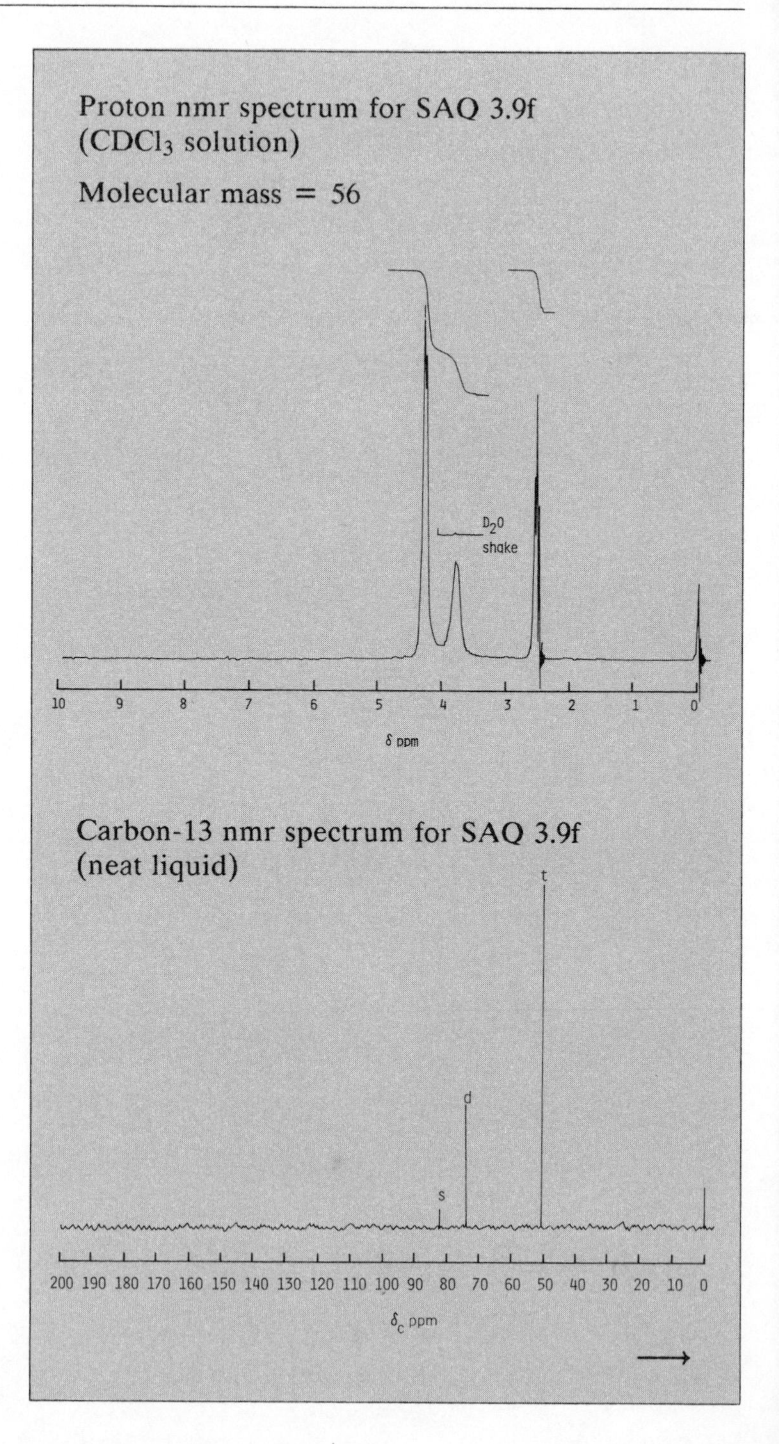
Proton nmr spectrum for SAQ 3.9f
($CDCl_3$ solution)
Molecular mass = 56
$D_2O$
shake
10
9
8
7
6
5
4
3
2
1
0
δ ppm
Carbon-13 nmr spectrum for SAQ 3.9f
(neat liquid)
t
d
s
200 190 180 170 160 150 140 130 120 110 100 90 80 70 60 50 40 30 20 10 0
$δ_c$ ppm
→

**SAQ 3.9f**

**Objectives of Part 3**

Now that you have completed Part 3 you should be able to:

- interpret, in terms of molecular structure, straightforward proton and carbon-13 spectra;
  - use proton and carbon-13 nmr correlation tables and charts for chemical shifts and coupling constants to identify different types of proton and carbon in molecular structures;
  - assign proton and carbon-13 nmr spectra for given molecular structures;
  - apply some general problem-solving heuristics to the interpretation of nmr spectra;
  - work logically with nmr data;
  - write a reasoned and detailed interpretation of an nmr spectrum.
- appreciate the use of problem-solving heuristics;
  - define a heuristic as a rule-of-thumb;
  - list six general heuristics;
- calculate the number of double bond equivalents for a given molecular formula containing C, H, N, O and a halogen;
- recognise that symmetry in a molecular structure can simplify nmr spectra yet make their interpretation difficult;
- recognise the enormous value of nmr to the area of structural chemistry and biochemistry.

Perhaps you hadn't realised nmr is one of the most powerful and widely employed spectroscopic techniques and that it has revolutionised structural chemistry since the early 1960s.

# 4. Signal Enhancement in NMR Spectroscopy

## Overview

In Part 1 we noted that integration of nmr signals could be used to derive quantitative data and we used integration ratios to help assign spectra in the qualitative problems of Part 3. Before we tackle some aspects of quantitative analysis in Part 5, in which we need to use accurate integrations, we have to discuss an important limitation of nmr spectroscopy, the inherent low sensitivity of the technique, and methods of improving or enhancing nmr signals.

In this part of the course, therefore, we are going to look more closely at the consequence of the Boltzmann distribution of spin state populations that you learnt about in Part 1. (Perhaps you should revise the first section of Part 1 if you are not sure about energy level populations). This will explain why nmr spectroscopy has inherently low sensitivity. Low sensitivity is obviously troublesome if we want to examine dilute spin systems such as in dilute solutions or as in the case of carbon-13 nuclei which are less sensitive and, of course, very much less abundant than protons.

We are also going to study sensitivity in terms of signal-to-noise ratio, and will see how to measure this using a standard solution of ethylbenzene. We shall then consider signal enhancement techniques which, with the aid of a computer, can be used to improve our spectra. Finally, we will spend some time on the most important technique for dilute spin systems—Fourier Transform (FT) nmr.

## 4.1. SIGNAL-TO-NOISE RATIO

As the concentration of material in solution decreases so the nmr signals become weaker. (This is, of course, an observation applicable to all instruments). The electronic design of the instrument allows us to increase the sensitivity, but at the same time electronic noise is introduced. Fig. 4.1a shows this for the $CH_2$ quartet of ethylbenzene. We can offset the increase in noise to some extent by switching in electronic filters as is shown in the Figure, but some signal distortion then starts to occur.

A useful analogy for this comes from hi-fi tape and record systems. As you may know record discs often have 'hiss' or high frequency noise associated with them. Most quality hi-fi amplifiers have a high frequency filter which can help remove such noise. Unfortunately, high frequency notes in the music are also lost or muted—a most unsatisfactory state of affairs if you are fond of violin concertos.

In nmr spectroscopy, as in other techniques, it is useful to define a signal-to-noise ratio, $S/N$, as a measure of how well the instrument can distinguish between signals and the electronic noise. The usual standard for CW proton nmr in such determinations is a 1% v/v solution of ethylbenzene in carbon tetrachloride. The spectrum of the methylene quartet is obtained under optimum spectrometer settings and the $S/N$ calculated according to Eq. 4.1a and Eq. 4.1b.

$$S/N = \frac{\text{Average Signal Amplitude}}{RMS\ \text{Noise}} \tag{4.1a}$$

$$RMS\ \text{Noise} = \frac{\text{Average Peak-to-Peak Noise}}{2.5} \tag{4.1b}$$

(Remember *RMS* is *root mean square*)

For a modern CW spectrometer an $S/N$ of 30 : 1 can be expected, but older instruments had much worse ratios in the region 5–10 : 1.

Fig. 4.1b shows a typical spectrum from a $S/N$ determination.

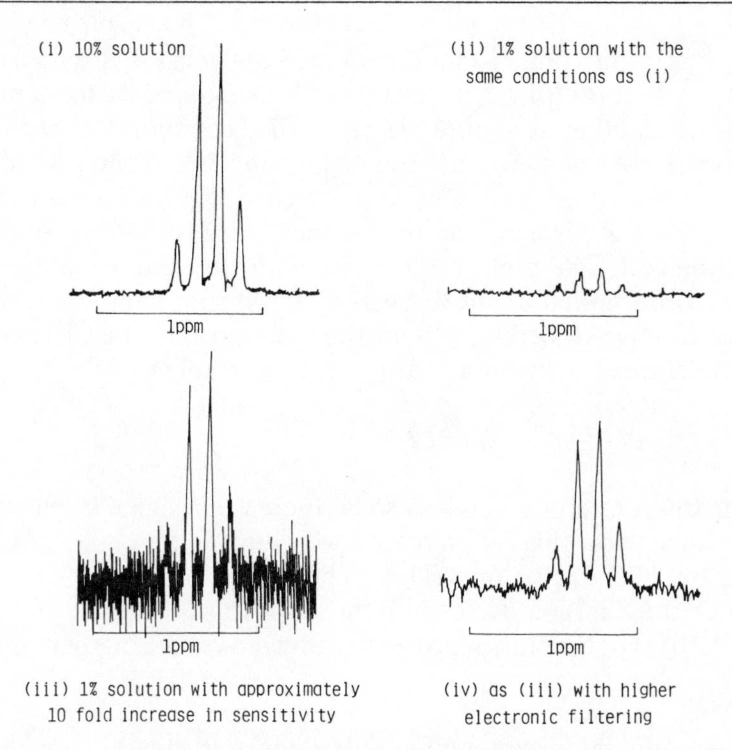


**Fig. 4.1a.** *Ethylbenzene quartet of 10% and 1% solutions at different instrument settings*

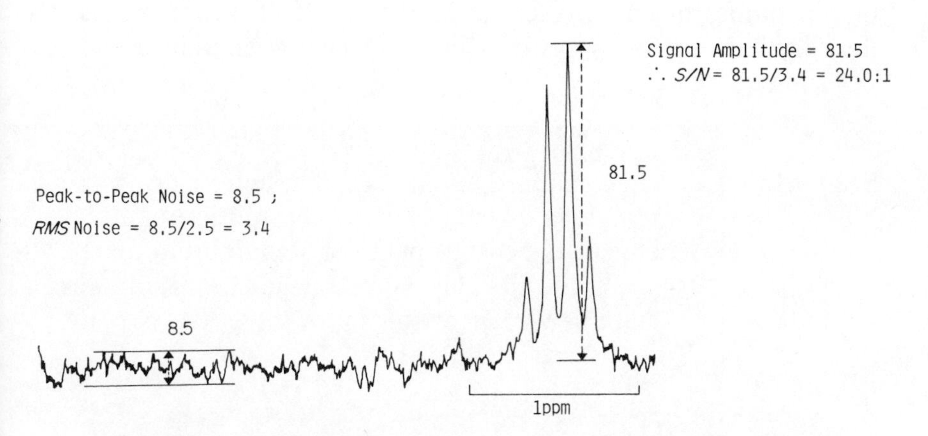


**Fig. 4.1b.** *S/N determination of the ethylbenzene quartet*

Two important points emerge immediately from this discussion. First, while it is difficult to relate $S/N$'s from one instrumental technique to another it should be clear that compared to techniques like mass spectrometry, atomic absorption spectrometry, and gas–liquid chromatography, CW nmr spectroscopy is a rather insensitive technique. The source of this lack of sensitivity is only partly instrumental. The real problem lies with the near equality of spin populations talked about in Section 1.1. If spin energy levels had a larger difference between them then there would be (according to the Boltzmann distribution, Eq. 1.1c) more spins

$$N_2/N_1 = 1 - \Delta E/kT \qquad (1.1c)$$

in the lower energy levels and so an increase in signal intensity upon resonance. Now this is one reason why high field nmr spectrometers (Section 2.2f) are so important. The higher the magnetic field the greater the energy difference between spin states (Section 1.1, and Eq. 1.1b). High field spectrometers have a much higher, intrinsic, $S/N$.

The second important point is that there is clearly a problem in obtaining an nmr spectrum from a dilute solution of spin-active nuclei. For example, with a 0.1 v/v solution of ethylbenzene we have difficulty finding a signal for the $CH_2$ protons (Fig. 4.1a) and even for a neat liquid sample we would have problems finding the carbon-13 resonances because, as well as being inherently less sensitive nuclei than protons, they are present as only 1.1% of all carbon atoms, The next sections explore ways in which $S/N$ can be enhanced.

**SAQ 4.1a**

When making an $S/N$ determination using the methylene quartet of a 1% v/v solution of ethylbenzene a peak-to-peak noise amplitude of 2.2 cm was found. The average signal amplitude was 17.5 cm. Which of the following is the $S/N$ for the instrument?

(*i*) 7.95 ⟶

**SAQ 4.1a (cont.)**

(*ii*) 19.9

(*iii*) 38.5

(*iv*) 0.39

**SAQ 4.1b**

Which of the following are reasons for the low sensitivity of nmr spectroscopy? Circle T (true) or F (false)

(*i*) There is a great deal of instrument electronic noise.

T / F

(*ii*) Spin–energy levels are too close to one another.

T / F

(*iii*) The sample might be too hot.

T / F

## 4.2. SIGNAL ENHANCEMENT TECHNIQUES

With the recent development of mini- and micro-computers, signal enhancement by mathematical and computational methods has become very important. Much of what follows in this section applies equally well to any instrumental signal, but is particularly pertinent for the low sensitivities encountered in nmr spectroscopy. As before in this Unit we are not going to be too worried about detailed explanations, but rather more concerned about the overall ideas most of which turn out to be quite easy to grasp.

The first thing we need to note if we have a computer to help us is that there has to be a device, an analogue–digital converter (ADC), between the nmr spectrometer and the computer. Such an *interface* allows the variable voltage output signal (the analogue signal) to be converted into a series of digital readings which the computer can handle. Fig. 4.2a shows these points. ADCs and computers can work so fast that several hundred readings can be taken per second. So our nmr spectrum ends up as a large array of digital values inside the computer's memory. We can now employ various types of *digital filtering* software to improve the $S/N$ before converting the digits back to an analogue signal and having this drawn out on a chart recorder.

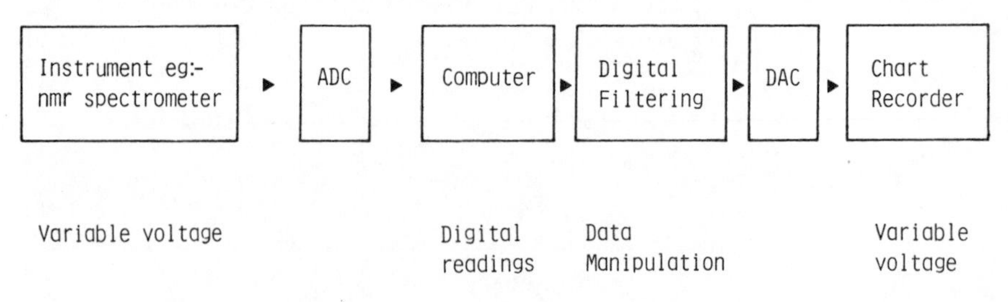


**Fig. 4.2a.** *Signal processing*

Three of the most common techniques of digital filtering are *boxcar averaging, ensemble averaging* (or CATing), and *weighted digital filtering* (or Savitsky–Golay smoothing). All of these computational methods reduce noise and are sometimes referred to as *smoothing routines*.

Boxcar averaging involves taking a lot of readings quickly of a slowly scanning analogue signal and averaging them. This average is then stored by the computer as a data point. The process is then repeated to get the next average point and so on. Fig. 4.2b shows this. So long as the nmr spectrum is being scanned slowly boxcar averaging can be used in *real time*, ie as the spectrum is being obtained. If $(S/N)_0$ is the signal to noise ratio of the data without a boxcar average then the improvement in $S/N$ is given by Eq. 4.2a.

$$S/N = \sqrt{n}(S/N)_0 \tag{4.2a}$$

where $n$ is the number of points averaged in each boxcar. Clearly the larger the value of $n$ the greater the improvement.

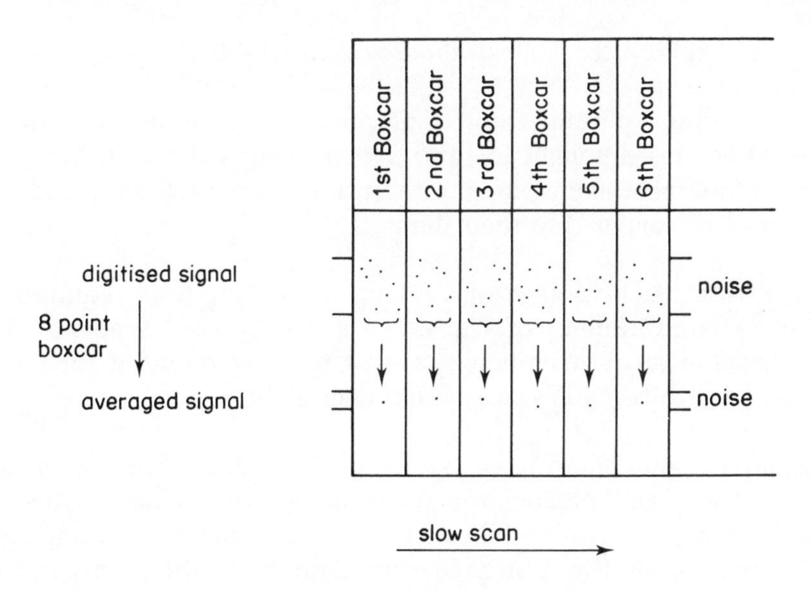


**Fig. 4.2b.** *Boxcar averaging*

Ensemble averaging (Fig. 4.2c) involves scanning the whole of the nmr spectrum a number of times, $n$, adding them together and dividing by $n$ to get an average spectrum. If each scan is obtained under exactly the same conditions then any real signals will add up on top of one another while any noise will average to zero. As with boxcar averaging the improvement in $S/N$ is, under ideal conditions, a factor of $\sqrt{n}$.

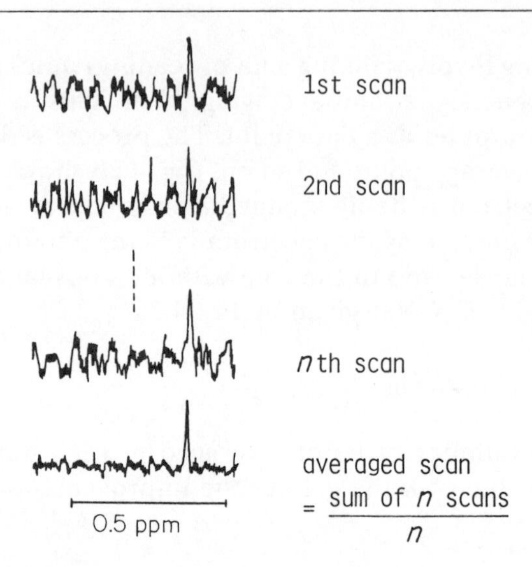


**Fig. 4.2c.** *Ensemble averaging or CATting*

At one time, specially built Computers of Average Transients (CATs) could be bought for nmr spectrometers. These CATs performed ensemble averaging, but their abilities were limited and the arrival of FT nmr led to their demise.

The obvious disadvantage of CATing is the long time required to obtain a large number of scans. If an average scan time was five minutes and an $S/N$ enhancement of ten was required then you would have to run 100 scans taking over eight hours.

Weighted digital filtering or Savitsky–Golay smoothing (after the physicist and mathematician who developed the definitive theory) is a bit more difficult to follow. In essence it involves averaging a set of points, say five, but giving more weight to the points nearer the centre, ie the second and fourth points would count most of all. The weighted average then replaces the central point and the calculation is repeated after moving one further point along the data. (Fig. 4.2d). Although this sounds a bit convoluted (and that is a pun for any mathematical cognoscenti amongst the readers) the method is ideally suited for computers.

Signal Amplitude

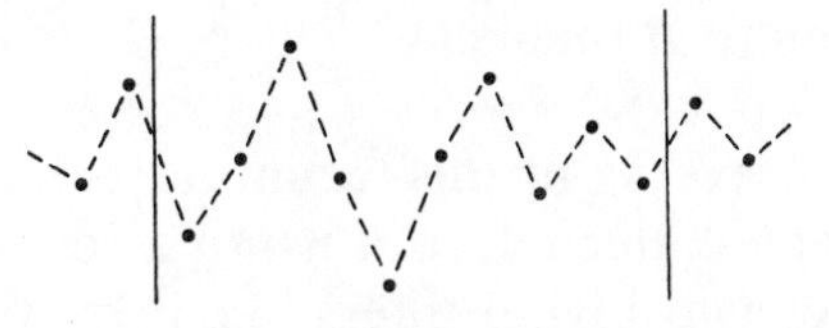

Data Point Number . . . . . . . . 1 2 3 4 5 6 7 8 9 10

↓

| Data Point Number | 1 | 2 | 3 | 4 | 5 | 6 | 7 | 8 | 9 | 10 |
|---|---|---|---|---|---|---|---|---|---|---|
| Signal Amplitude | 115 | 122 | 133 | 120 | 110 | 122 | 130 | 118 | 125 | 119 |
| Smoothed Signal Amplitude* | 121 | 123 | 128 | 120 | 114 | 121 | 125 | 124 | 120 | 124 |

↓

Smoothed Signal Amplitude

Data Point Number . . . . . . . . 1 2 3 4 5 6 7 8 9 10

*Smoothing function:

$$\text{data point } n = \frac{\begin{array}{l} -3 \times \text{data point } (n-2) \\ +12 \times \text{data point } (n-1) \\ +17 \times \text{data point } (n) \\ +12 \times \text{data point } (n+1) \\ -3 \times \text{data point } (n+2) \end{array}}{\text{divided by } 35}$$

**Fig. 4.2d.** *Weighted digital filtering*

The coefficients −3, +12 and +17 are known as convolution coefficients and together with the normalising factor, 35, are derived from mathematical theory.

Unlike boxcar averaging this technique can only be used after all the data have been collected, so it is not a real-time method. However, like the other two digital filters, Savitsky–Golay smoothing should give, under ideal circumstances, an *S/N* improvement of $\sqrt{n}$ where $n$ is the number of points in the set (five in our example).

The most effective *S/N* enhancement comes from using these digital filters in combination. Not only is the time required for CATing reduced, but also the different techniques complement one another. Fig. 4.2e shows a set of spectra of the ethylbenzene quartet obtained from a 1% solution under routine spectrometer conditions. Firstly, there is the 'raw' spectrum without any enhancement. The second of the set is equivalent to the first, but with a 16 point boxcar average. There is a clear reduction. The third part of the Figure shows the effect of a 16-fold CATing procedure on the boxcar spectrum and the final spectrum is the result of a 17 point Savitsky–Golay smooth on the CAT spectrum. The signal enhancement is quite obvious.

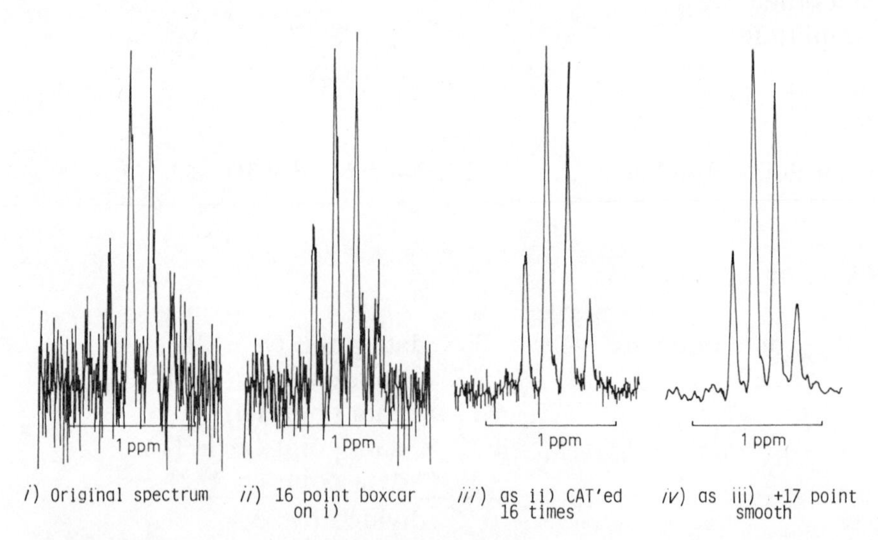


**Fig. 4.2e.** *Signal enhancement on the ethylbenzene quartet (1% solution)*

Fig. 4.2f shows these techniques in action in a real problem. Propranolol, a drug used in the treatment of heart disease, is only slightly water soluble. Consequently, its nmr spectrum is rather weak and noisy. In order to establish chemical shifts and coupling constants in one particular region signal enhancement had to be done. The third section of the Figure shows the rather dramatic enhancement from a single 50 point boxcar average followed by a 17 point smooth. There was *no* CATing needed in this case.

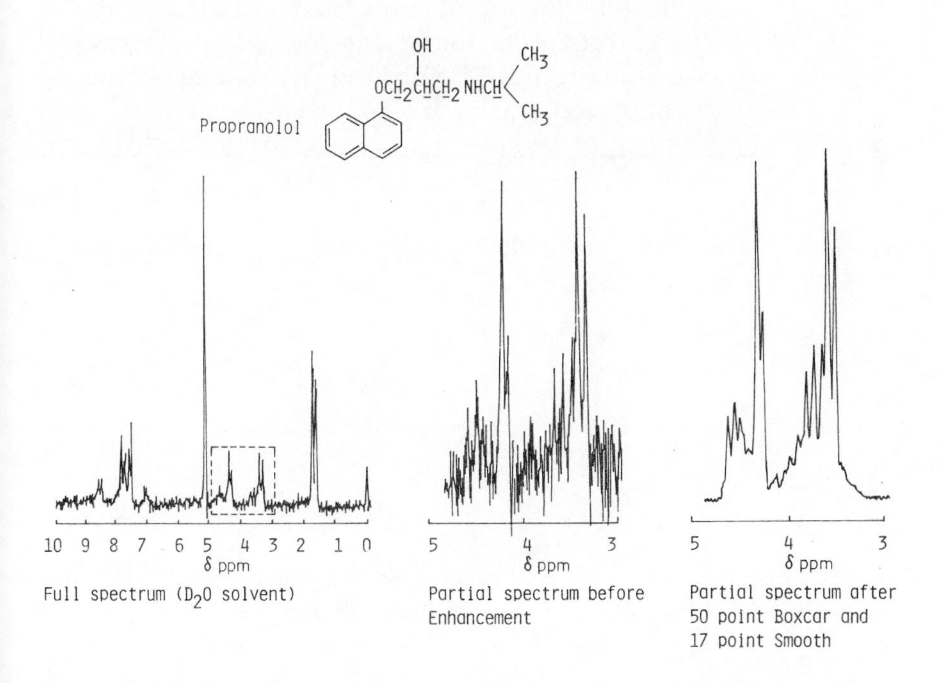


**Fig. 4.2f.** *Signal enhancement of the nmr spectrum of propranolol*

While these examples show the potential of computer enhancement the predicted *S/N* improvement is rarely achieved. For both the boxcar and ensemble averaging method the results are dependent on the sampling rate, ie how often the signal is converted into a digit. Sometimes, if the sampling rate is too high noise can actually be introduced into the data. There are two other disadvantages with these techniques. First, with CATing, the nmr spectrum must be quite stable and reproducible otherwise signal broadening results. Secondly, loss of resolution also occurs with Savitsky–Golay smoothing because of the nature of the mathematics. Nevertheless, empirical use of these techniques can give useful results.

**SAQ 4.2a**

You have decided to try some signal enhancement on your nmr signals. List and arrange in the correct order four pieces of equipment you are going to need before the signals emerge on a chart recorder.

**SAQ 4.2b**

Match the following statements with the terms, boxcar averaging, ensemble averaging, and weighted digital filtering.

(*i*) Multiscan, real-time spectral averaging.

(*ii*) Signal-to-noise enhancement by a factor of $\sqrt{n}$.

(*iii*) Post-scan data smoothing.

(*iv*) Rapid multipoint averaging of slow scan signals.

**SAQ 4.2c**

You estimate that $S/N$ enhancement by a factor of sixty-four is necessary to get useful nmr data from the spectrum of a particular sample. Which *one* of the following combinations of digital filtering techniques is going to be the most useful?

(*i*) A 16 point boxcar on each of a four-fold ensemble average followed by a 16 point Savitsky–Golay treatment.

(*ii*) A 32 point boxcar with a 4 scan CAT followed by a 32 point weighted filter.

(*iii*) A 16 point boxcar with 32 ensemble averages, then an 8 point Savitsky–Golay smooth.

(*iv*) A 128 point boxcar with a 4 scan CAT followed by an 8 point post scan smooth.

## 4.3. FOURIER TRANSFORMATION

Notwithstanding the digital techniques of the last section, by far the most important signal enhancement method uses a mathematical device called Fourier transformation (FT). In FT nmr spectroscopy the sample is given a powerful *pulse* of radiofrequency radiation. This pulse of radiation contains a broad band of frequencies (Fig. 4.3a) and it causes all the spin-active nuclei to resonate at once at their Larmor frequencies. (In CW spectroscopy a frequency range is slowly scanned at low power, bringing the spin-active nuclei into resonance each at its own frequency). The detector system in the spectrometer senses the change in magnetisation of the sample and the decay of the magnetisation with respect to time. This decay is called the *free induction decay* (FID). As a number of frequencies are involved the FID does not have a simple pattern, but is composed of a complex set of interfering wave forms along with a great deal of noise. Fig. 4.3b shows a computer generated FID for the protons of an ethyl group ($CH_3CH_2-$). Although there is no noise in this example the spectrum is still rather complicated.

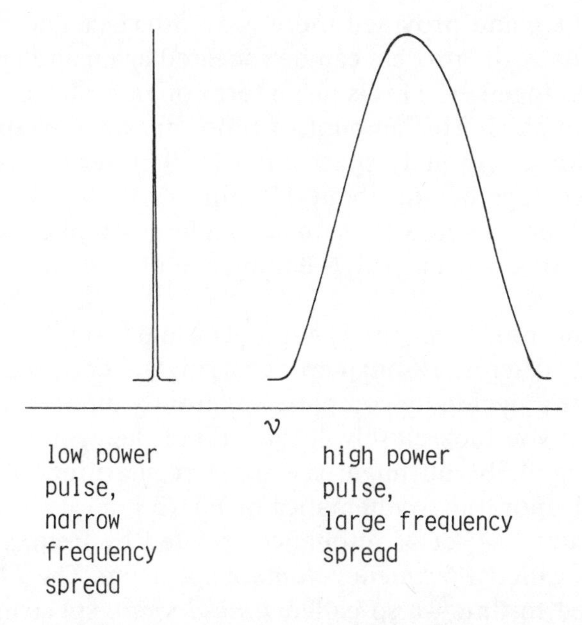


**Fig. 4.3a.** *Frequency spread of radiation pulses*

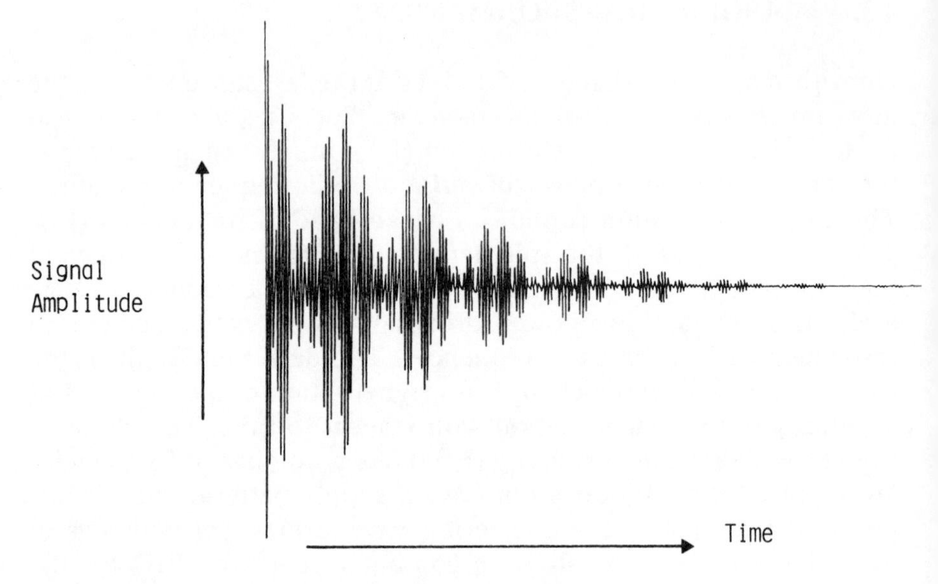


**Fig. 4.3b.** *Free induction decay for the protons in an ethyl group* $(CH_3CH_2-)$

The pulse and the subsequent detection of the FID can take as little as a second so, and provided there is an interface and computer to record the data, the process can be repeated again and again and the FID's added together. This is just a form of ensemble averaging, but it takes a much shorter amount of time that for CW spectroscopy. Typically, for a carbon-13 spectrum a 1000 pulses can be given and FID's added together in about 15 minutes (how long would that take in CW spectroscopy?). With such a large number of data being collected in this way large $S/N$ enhancements are soon achieved.

However, our nmr spectrum is not in its usual form. Instead of being a number of discrete resonances occurring at very precise frequencies we have a decidedly complex, apparently uninterpretable FID. (Although if you look closely at the FID of the protons in the ethyl group in Fig. 4.3b you might see some regularities.) At this point we can call upon the mathematics of FT to help us out. A normal nmr spectrum is a set of resonances related by frequency—such a spectrum is called a *frequency-domain* spectrum. The FID is a spectrum related to time—a so-called *time-domain* spectrum. Now FT mathematics is a method of changing a time-domain spectrum into a frequency-domain spectrum (or vice versa). For completeness sake the two mathematical expressions which transform frequency- and time-domains are shown below. $F(\nu)$ is our frequency spectrum and $f(t)$ is the time spectrum.

$$F(\nu) = \int_{-\infty}^{+\infty} f(t)\, e^{-i(2\pi)\nu t}\, dt \quad (4.3a)$$

$$f(t) = \int_{-\infty}^{+\infty} F(\nu)\, e^{i(2\pi)\nu t}\, 2\pi d\nu \quad (4.3b)$$

The object for you here is not to learn these equations, but to appreciate the complexity of the mathematics used. A computer is essential to tackle the problem.

Fig. 4.3c shows the relationship between a very simple FID, and its corresponding single nmr frequency signal. Note how all the information in the frequency-domain spectrum is contained in the time-domain version, chemical shift, area, and relaxation time.

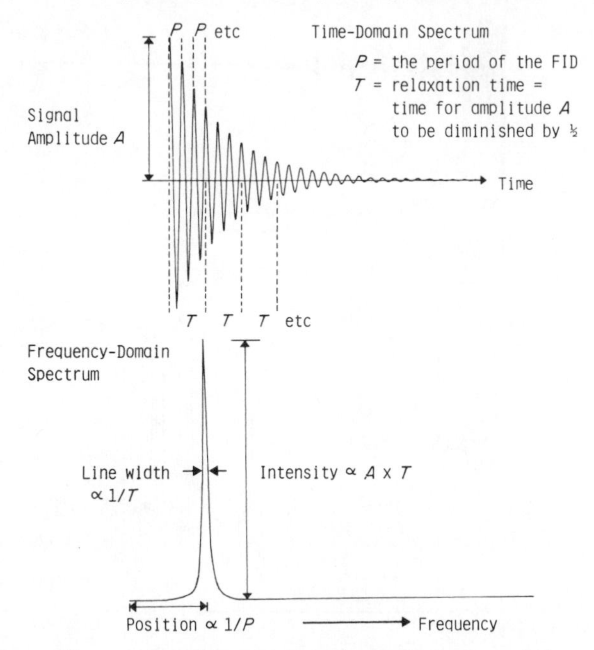


**Fig. 4.3c.** *Time- and frequency-domain spectra for a single resonance*

So Fourier transformation of an FID gives us the more familiar nmr spectrum.

Fig. 4.3d shows the time-and frequency-domain nmr spectra of the protons in pyridine ($C_5H_5N$). The frequency spectrum is extremely complex and its interpretation is well beyond this course. However, all that information is contained in the time-domain spectrum.

Fourier Transform nmr is a very complex technique and the foregoing description only scratches the surface of what is involved. Nevertheless, we now have enough knowledge to understand the main idea and to appreciate how nmr spectra of dilute spin systems like carbon-13 can be obtained.

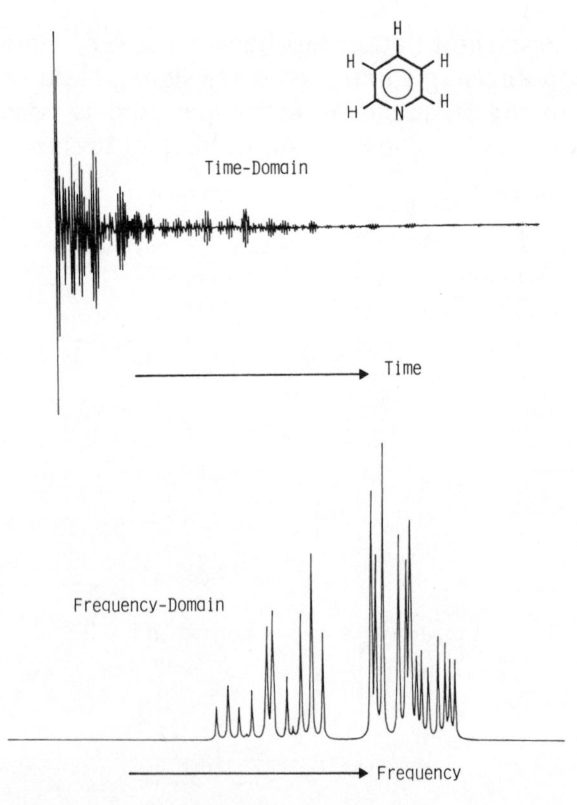


**Fig. 4.3d.** *Time- and frequency-domain nmr spectra for the five protons in pyridine*

**SAQ 4.3a**

From the list of words/phrases fill in the blanks in the following brief description of FT nmr.

FT nmr uses ... (1) ... of radiofrequency radiation to excite spin-active nuclei. The decay of magnetisation with time, the ... (2) ... is recorded and stored in a computer. The procedure is repeated as often as necessary and the resulting ... (3) ... ... (4) ... transformed to an ... (5) ... .

**Words/phrases:**

time-domain spectrum, frequency-domain spectrum, pulses, free induction decay, noise averaged.

**SAQ 4.3b**

Are the following statements about free induction decay spectra correct or incorrect? Circle T (true) or F (false).

(*i*) FID's contain all the nmr information except coupling constants.

T / F

(*ii*) FID's cannot be interpreted because they are always too noisy.

T / F

(*iii*) Fourier transformation of an FID involves simple mathematics.

T / F

## 4.4. SUMMARY OF PART 4

In this part of the Unit we have seen a number of ways that can be used to enhance the signal from the nmr spectra of dilute spin-active systems like carbon-13 or dilute solutions of protons. Using the idea of signal-to-noise ratio, $S/N$, we explored the noise averaging techniques of boxcar, ensemble, and weighted digital filtering as methods of signal enhancement. We noted that a computer interfaced to the nmr instrument was essential for treating the data. While such data smoothing techniques have their place, the technique of pulsed nmr with Fourier transformation provided the most powerful method of signal enhancement, via ensemble filtering. A free induction decay spectrum could be obtained in less than a second following a pulse, and hence a large number of FIDs could be added together over a relatively short period of time to give an averaged time-domain spectrum. Upon Fourier transformation the more familiar frequency-domain spectrum is then obtained.

**SAQ 4.4a**

> Make a list of at least six ideas, techniques, or things that have been referred to in Part 4 by their initials, eg *RMS*—Root Mean Square. (And you cannot count that one!)

**SAQ 4.4b**

Which of the following statements about electronic noise are correct? Circle T (true) or F (false).

(*i*) Noise can never be completely eliminated from an nmr spectrum.

T / F

(*ii*) Useful information can be hidden in noise.

T / F

(*iii*) Noise deforms nmr signals.

T / F

(*iv*) Noise tends to average to zero if nmr signals are repetitively scanned and accumulated.

T / F

**Objectives of Part 4**

Now that you have completed Part 4 you should be able to:

- explain the inherent low sensitivity of the nmr experiment in terms of the Boltzman distribution of spin states populations;

- appreciate the concept of signal-to-noise ratio ($S/N$) in nmr spectroscopy;

  - define $S/N$ in terms of average signal amplitude and $RMS$ noise;

  - calculate the $S/N$ for a given nmr spectrum;

  - state that for proton and carbon-13 spectra a 1% v/v solution of ethylbenzene is used as a standard for $S/N$ determination;

- recognise the importance of signal enhancement techniques for obtaining nmr spectra of dilute spin-systems;

  - explain the purpose of a computer interfaced to an nmr spectrometer;

  - sketch the arrangement of the components of a computer–nmr system;

  - list three enhancement techniques as boxcar averaging, ensemble averaging (CATing) and digital filtering;

  - explain the principle of boxcar averaging in terms of rapid multipoint averaging of slow scan signals;

  - explain the principle of ensemble averaging in terms of multiscan, real-time averaging;

  - explain the principle of digital filtering in terms of post-scan data smoothing;

  - appreciate that the enhancement obtained is theoretically $\sqrt{n}$ where $n$ is the number of data points (or scans) involved in the enhancement function;

  - list the advantages/disadvantages of these enhancement techniques;

- appreciate the importance of Fourier Transform (pulse) nmr spectroscopy as an $S/N$ enhancement technique;

  - explain the basic difference between FT and CW nmr spectroscopy;

  - compare multipulse FT nmr with CW ensemble averaging in terms of time requirements;

  - sketch a free induction decay;

  - explain the terms frequency-domain and time-domain spectrum;

  - state that frequency-domain and time-domain spectra are related by a complex mathematical function called a Fourier transform;

  - show via sketches that the information in a time-domain spectrum is contained in a frequency-domain spectrum;

  - appreciate the use of computers to help carry out FT nmr spectroscopy.

# 5. Quantitative Applications of NMR Spectroscopy

## Overview

The final part of this course is devoted to quantitative problems that can be tackled via the use of integration of nmr signals. We are going to find that there is a wide range of problems, from working out the relative amounts of drugs in a mixture to the determination of molecular masses of polymers, that can be tackled by nmr spectroscopy.

By the end of this Part you should have a good grasp of the fairly easy ideas that are involved. Beware, however, there are some tricky points to understand and you will need to have a fair knowledge of the structures of organic molecules. Nevertheless, you should be able to describe several examples of quantitative nmr and carry out the calculations needed for these and similar examples. You should find the SAQs particularly helpful in this respect.

You should also bear in mind that in a course of this nature we can only scratch the surface of the subject and there are additional important applications that we have not covered.

We first used integrations in Part 3, in qualitative analysis, and we are now going to concentrate on their use in real problems of quantitative analysis. But first, there are three general points to bear in mind.

I shall be using these specific applications to build up some general ideas about quantitative nmr analysis. Yet as a practising analyst I prefer to work out each case from first principles rather than rely on predetermined, sometimes complex, equations. Luckily with integrations and nmr spectroscopy such deductions are usually quite easy.

Second, the following examples are limited to proton and fluorine-19 nmr. Carbon-13 spectra do not, in general, give reliable integrations. This is due to relaxation and other effects.

Finally, as an ongoing activity you should think about alternative ways of tackling the following problems. It is necessary for an analytical laboratory to ensure the accuracy of its work, and one way of doing this is to carry out analysis using a totally different technique.

## 5.1. THE CASE OF THE DIFFICULT DRUG

APC tablets are an analgesic for bodily aches and pains (although they are no longer available in the UK). They contain a mixture of three drugs, aspirin, phenacetin, and codeine, whose initials give the tablets their name. The structure of aspirin and phenacetin are shown on Fig. 5.1a. Codeine has a complex alkaloid structure and its nmr spectrum, fortunately, does not obscure the region of interest in this example.

Proton nmr provides a quick, easy, and reliable method for determining the relative amounts of aspirin to phenacetin as a molar ratio in these tablets. Fig. 5.1a shows the proton nmr spectrum obtained from the $CDCl_3$ extract of a crushed APC tablet. Although the spectrum appears rather confusing we can look at the two sharp singlet resonances at $\delta = 2.1$ ppm and 2.3 ppm. They come from the acetyl methyl groups in phenacetin and aspirin respectively. So by measuring the integration of both peaks we can obtain the molar ratio of aspirin and phenacetin.

From the integrations on the spectrum, the measured step heights average to 1 : 1. Thus aspirin and phenacetin are present in equal

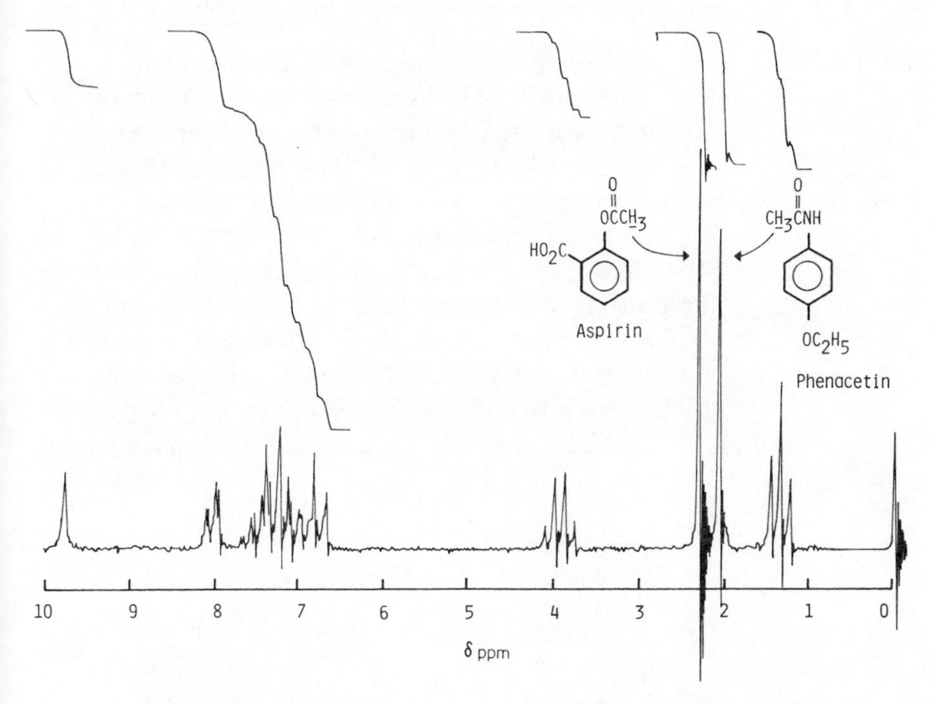


**Fig. 5.1a.** *Proton nmr spectrum of an APC tablet*

molar amounts. Normally when we use step curves in quantitative work we obtain 5 or 7 and average them. However, for clarity on the spectrum I have shown only one curve.

Such an analysis could be carried out in less than twenty minutes by an experienced operator; far quicker than any 'wet' chemical analysis.

**SAQ 5.1a**

An inexperienced nmr operator decided to analyse a commercial APC tablet for the relative quantities of aspirin and phenacetin. Not having a mortar and pestle to hand he crushed the tablet on a filter paper with a spatula and having placed the resulting lumps in a small flask he extracted them with carbon tetrachloride. Then he poured the carbon tetrachloride solution directly into an nmr tube. Nmr analysis gave him a ratio of 1 : 2 aspirin to phenacetin. List at least three errors that the operator made.

## 5.2. THE CASE OF THE SINGULAR SPIRIT

In our first case, although the structures involved were complex, the analysis turned out to be easy because there were two separate readily identifiable resonances for integration. In this second case, the determination of the percentage by volume (% v/v) of ethanol in alcohol/water mixtures, (eg beers, wines, and spirits) the two structures—ethanol and water—are simple and give well separated resonances, yet the analysis has a couple of complexities.

Fig. 5.2a shows the proton nmr spectrum of a single malt whisky. Luckily, only the signals for the hydroxyl protons at $\delta$ = 5.0 ppm and the ethanol protons at 3.8 ppm ($CH_2$) and 1.2 ppm ($CH_3$) are visible. All the other minor colouring and flavouring constituents are of such low concentration so as to be nmr invisible. As usual, integrations can be obtained and the numbers on the spectrum show the values obtained from seven digital integrations of the resonances. We can use either the average values of the readings or their totals. Using averages, errors functions can be calculated and you can do that as an extra exercise.

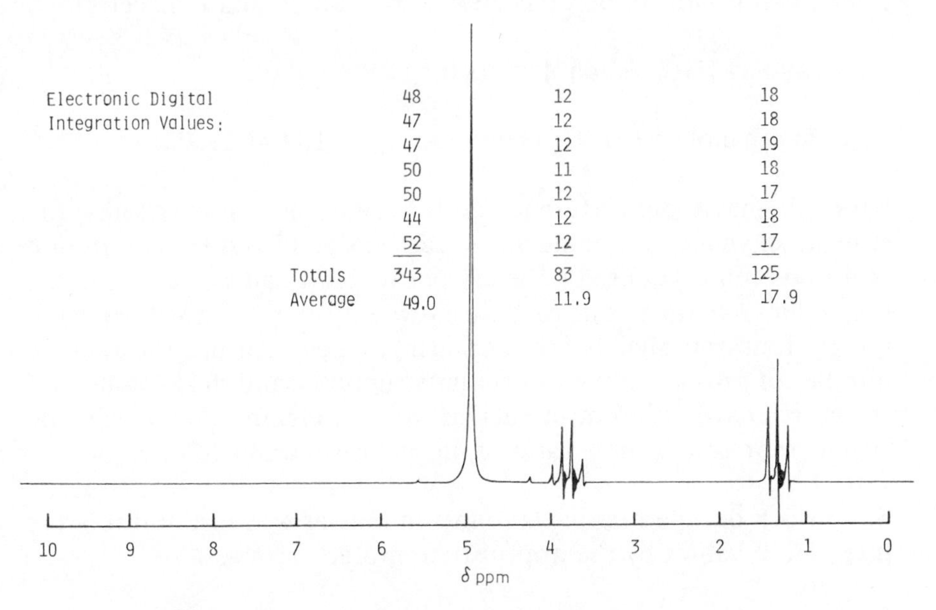


**Fig. 5.2a.** *Proton spectrum of a neat whisky*

You should see immediately that the ratio of the $CH_3$ to OH integrations does *not* give the ratio of ethanol to water, nor does the $CH_2$ to OH integration ratio. What you have to realise is that these signals come from groups having different numbers of protons associated with them. Now there are three protons on the methyl group, two on the methylene group (and the ratio of those two signals is 3:2), and it is tempting to think that there are two protons (from $H_2O$) associated with the hydroxyl signal at 5.0 ppm. However, that signal also contains the OH resonance for the ethanol hydroxyl proton.

Once this problem is recognised it is easily overcome. If the $CH_3$ and $CH_2$ protons have integrations of 11.9 and 17.9 then one proton in ethanol (ie the OH proton) will have an integration of 6.0 (to one decimal place). So we subtract this value from the total value of the OH signal to give the integration for the water in the sample. Then we can proceed with the calculation as follows:

Integration for $H_2O$ = 49.0 − 6.0 = 43.0;

Integration for $CH_2$ = 11.9

(*Note*: two proton in both species, so we can compare directly).

$\therefore$ ratio of $H_2O$ : ethanol = 43.0 : 11.9

$\therefore$ % ethanol = (11.9/(43.0 + 11.9)) × 100 = 21.7%

Now this may seem a strange result because as you may know the strength of whisky as laid down by Customs and Excise is a minimum of 40% volume—quoted on every bottle. How can we reconcile the two values? It turns out to be easy when we remember what the integrations/nmr signals are measuring. They measure the relative number of protons in the molecular structures, and those molecular structures have different molecular masses. Or, in other words, the signals represent a *mole* ratio of, in this case, water to ethanol.

So to work out the relative amounts in masses we must multiply the integration values by the appropriate molecular mass:

ratio of $H_2O$ : ethanol (mass : mass)

$= 43.0 \times 18 : 11.9 \times 46$

$= 774 : 547.4$

$\therefore$ % ethanol

$= (547.4/(774.0 + 547.4)) \times 100 = 41.4\%$ w/w (by mass)

Well, that ratio is a bit more like it, but we have yet to convert to volumes, the density of ethanol being 0.96 g $cm^{-3}$.

Thus the ratio of $H_2O$ : ethanol (volume : volume)

$= 774.0/1 : 547.4/0.96 = 774.0 : 570.2$

$\therefore$ % ethanol $= (570.2/(774.0 + 570.2)) \times 100 = 42.4\%$ v/v

In fact this particular malt whisky quotes 43% by volume on the bottle.

Our whisky is the expected strength after all.

This example has highlighted the need to compare equivalent signals (we did this in the APC example because we compared two methyl signals), and the use of molecular masses to obtain mass : mass ratios. The method can be applied to wines and beers and indeed to other types of water/ethanol mixtures, although with beer having a low alcohol concentration (about 4% alcohol) measurements have to be made very carefully to avoid signal-to-noise problems inherent in comparing a very large signal (from $H_2O$) with a very small signal (from $CH_3CH_2OH$). (See the next Section for another way of tackling this type of problem).

**SAQ 5.2a**

A sample of chateau-bottled claret (a red Bordeaux wine) analysed by nmr spectroscopy gave integration values of 876.0 for the hydroxyl resonance and 52.0 for the methylene resonance of the ethanol. Which of the following represents the strength of the wine?

(*i*) 6.57 : 1 water : ethanol mass ratio

(*ii*) 14.0% ethanol by volume

(*iii*) 16.8 : 1 water : ethanol mole ratio

(*iv*) 6.39 : 1 water : ethanol volume ratio

## 5.3. THE CASE OF THE WONKY WAX

In the previous case there was a large integration for each of the two compounds in the mixture, but that need not always be the case. There may be only a small amount of one component or the signals used for comparison may be small. In these cases we can often use a favourite analytical procedure—the method of standard additions—to obtain a reliable result.

In this example a waxy hydrocarbon was known to be a mixture of polyethylene $(-CH_2CH_2-)_n$ and stearic acid ($C_{18}H_{36}O_2$). Our analysis must find the percentage by weight of stearic acid. The spectrum of the mixture is shown in Fig. 5.3a. The weak resonance at 2.2 ppm comes solely from the $CH_2$ adjacent to the carboxylic acid function in stearic acid. The intense resonance at 1.2 ppm is that for the polyethylene protons and the other stearic acid methylene protons. Clearly, the comparison of such a small resonance to the large one is going to be unreliable.

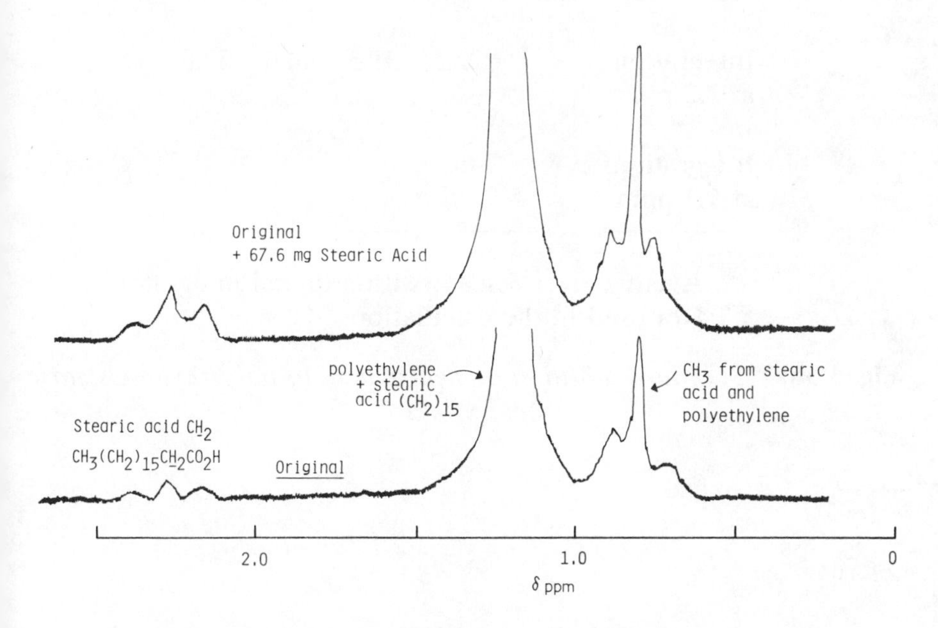


**Fig. 5.3a.** *Proton nmr spectra of stearic acid/polyethylene mixtures*

The answer here is to measure the integration of the acid $CH_2$ peak at 2.2 ppm as best as can be done from a known amount of the wax and then add a known quantity of stearic acid, remeasure, repeat the addition, and so on. In this way the signals build up and we can plot a graph of integration value for the low field methylene against amount of stearic acid added. (Fig. 5.3b and 5.3c). By extrapolation the amount of stearic acid in the original wax can be found. In this case the mixture is about 36% by weight stearic acid, 64% polyethylene. The graph shows 43 mg stearic acid is present in 120 mg of sample (ie 36%).

You should note carefully that this result did not depend on the integration of the large peak in the spectrum.

| Sample (Original mass 120 mg) | 1 | 2 | 3 | 4 |
|---|---|---|---|---|
| Mass of Stearic Acid added (mg) | 0 | 18.3 | 41.9 | 67.6 |
| Integration* at 2.2 ppm | 22.5 | 31.5 | 44.0 | 57.0 |
| Integration** at 1.1 ppm | 748 | | | |

* Average over 5 scans with a digital integrator
** Not used in the calculation

**Fig. 5.3b.** *Standard addition of stearic acid to polyethylene/stearic acid mixture*

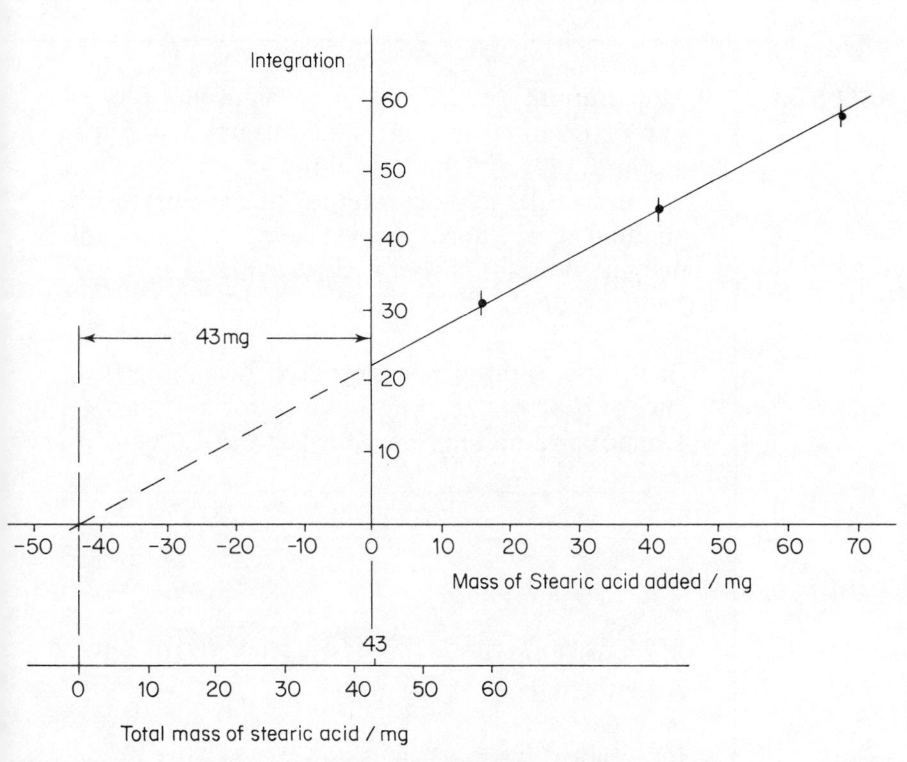


**Fig. 5.3c.** *Graph of standard addition of stearic acid to stearic acid/polyethylene mixture*

**SAQ 5.3a**

The amount of alcohol in a sample of beer was determined by nmr spectroscopy using the method of standard additions. From the following results decide whether the brewer, who claimed that there was at least 4% alcohol present, was a charlatan or a paragon of business practice.

Each nmr sample was prepared by taking 0.50 $cm^3$ of the beer adding the appropriate mass of ethanol and making up with $H_2O$ to 1.00 $cm^3$.

| | Sample | | | | |
|---|---|---|---|---|---|
| | 1 | 2 | 3 | 4 | 5 |
| Mass of Ethanol Added (mg) | 0.0 | 10.0 | 20.0 | 50.0 | 100.0 |
| Integration for $CH_3$ Resonance | 5 | 11 | 16 | 30 | 54 |

## 5.4. THE CASE OF THE PALTRY POLYMER

All the previous examples have concerned the analysis of mixtures. This example looks at one way in which we can obtain molecular masses of materials; in this case a fairly low molecular mass polyethylene. Remember that the value we shall get will not be a precise figure for two reasons: firstly, we are dealing with a polymer that will have a spread of molecular masses, and secondly, we are going to have to compare large values with small ones and for nmr integrations this is an important source of error.

Nevertheless, Fig. 5.4a shows the spectrum of a low density, low molecular mass polyethylene. As in the previous case we see the sharp resonance of the methylene protons at 1.2 ppm (but with no interfering stearic acid) and a distorted triplet of the terminal methyl groups at 0.9 ppm.

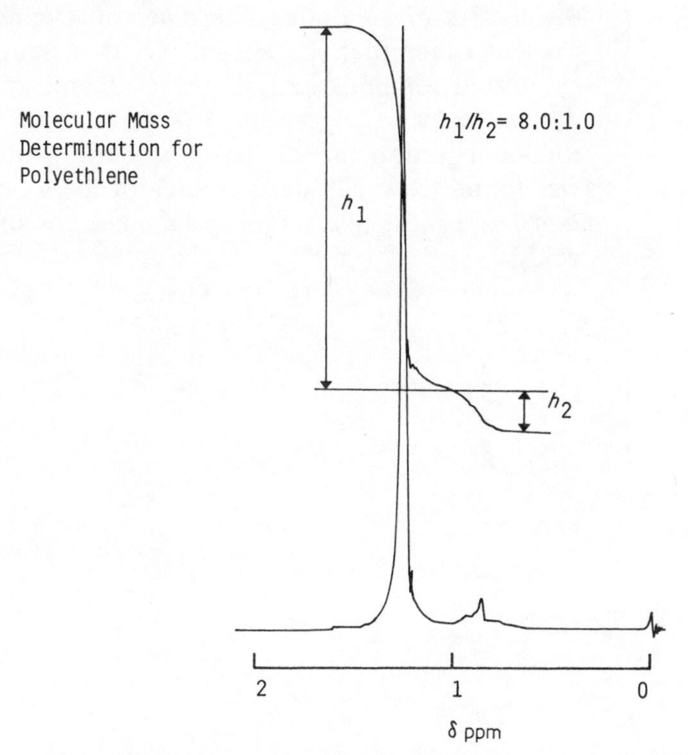


**Fig. 5.4a.** *Molecular mass determination for polyethylene,* $CH_3(CH_2)_nCH_3$

Careful integration gives a ratio of 8.0 : 1.0 for the methylene to methyl ratio. Now remembering the lesson of comparing like with like from an earlier example you should realise that the methyl resonance has six protons associated with it (there are two terminal methyls). Thus proton for proton the ratio becomes 8.0 × 6 : 1.0 ie 48.0 : 1.0. This means there are about 48 protons in the average methylene chain or 24 methylenes. The formula is thus:

$$CH_3(CH_2)_{24}CH_3$$

and the average molecular mass will be about 366.

**SAQ 5.4a**

Oleic acid is a long chain carboxylic acid which contains one —CH=CH— group as part of the hydrocarbon chain. There are no elements present other than C, H, and O. In a sample submitted for nmr analysis these alkenic resonances had an integration of 14 units while the total integration for all the rest of the protons was found to be 222 units. Which of the following is the most likely molecular mass for oleic acid?

(*i*) 256

(*ii*) 254

(*iii*) 282

(*iv*) 255

**SAQ 5.4b**

Ethylene/vinyl acetate copolymers (EVAs) are used extensively in the plastic industry and have the general structure and composition shown below.

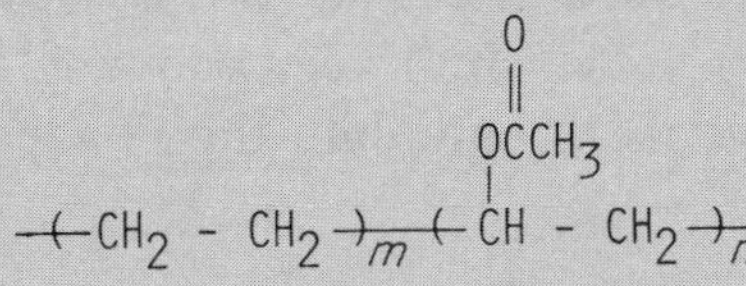

$$-\!\!\left(CH_2 - CH_2\right)_m\!\!\left(\underset{\substack{|\\ \mathrm{OCCH_3}\\ \|\\ \mathrm{O}}}{CH} - CH_2\right)_n$$

ethylene units    vinyl acetate units

The acetyl protons and the methylene protons give separate resonances whose areas can be measured by integration. (The methine proton also has a separate signal, but this does not affect the calculations). In one particular sample of EVA the integration ratio of methylene protons to acetyl protons was found to be 3 : 1. Which of the following is the percentage by weight of vinyl acetate in the sample?

(*i*) 84.3%

(*ii*) 15.7%

(*iii*) 75.4%

(*iv*) 25.0%

## 5.5. THE CASE OF THE DOUBTFUL DEUTERATION

Very often in quantitative nmr work the sample with which we are working will not have a peak which is free of interference and be associated with a known number of hydrogen atoms. To proceed with such a sample we can add to it a known amount of an *internal* standard. And, as we shall see, knowing the concentration of the standard allows us to calculate the concentration of the unknown material.

Of course, the internal standard must have a separate and readily identifiable resonance for integration. A good standard should have several other qualities, eg it should be sufficiently soluble in the appropriate solvent, should not react with the compound being studied, should not evaporate or decompose, and should have, ideally, only one single sharp signal. Fig. 5.5a lists a few compounds that can be used. There are many more possibilities.

| Compound | Useful Resonance ($\delta$ ppm) |
|---|---|
| Cyclohexane | 1.4 |
| Sodium ethanoate | 1.9 |
| Ethanoic acid | 2.1 |
| 1,1,1-trichloroethane | 2.7 |
| Methanol | 3.5 |
| Iodoform ($CHI_3$) | 4.9 |
| Dichloromethane | 5.2 |
| Benzene | 7.4 |

**Fig. 5.5a.** *Some compounds used as internal standards in quantitative nmr*

A few moments study should convince you that there is no ideal standard. All have one or more disadvantages. For example: methanol is volatile and has two resonances (one of which is broad), benzene is poisonous, ethanoic acid might be quite reactive and so on.

In this analysis we are going to investigate the purity of a sample of methanol-$d_4$ and to determine the mole percentage of protons in it.

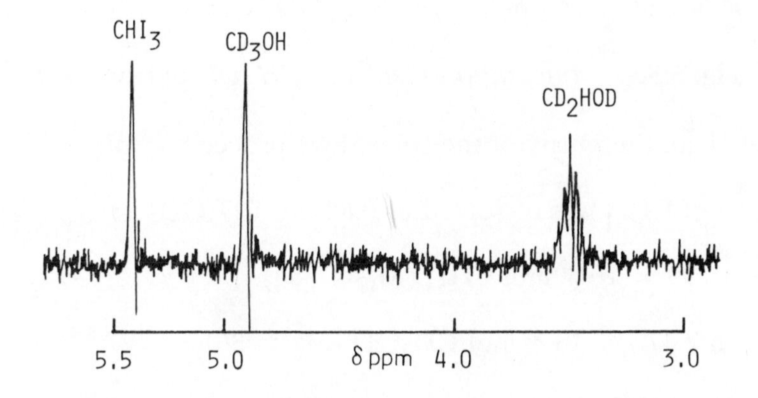


**Fig. 5.5b.** *The proton nmr spectrum of methanol-$d_4$ + iodoform*

As you probably have gathered from the previous Parts deuterated solvents like $CDCl_3$ and $CD_3OD$ are never 100% deuterated, there are always residual protons in the solvent albeit at a low level. For $CD_3OD$ there are two proton containing impurities, $CD_3OH$ and $CD_2HOD$. These two compounds give rise to signals at about $\delta$ = 4.9 ppm and 3.5 ppm respectively. The $CD_2H$ resonance shows the expected coupling to deuterium. In the sample under consideration (Fig. 5.5b) measurement of the integrations of these two peaks would tell us the ratio of $CD_3OH$ to $CD_2HOD$ as contaminants, but could not tell us the actual number of moles of each. So we add a known amount of iodoform which gives the third peak in Fig. 5.5b at 5.4 ppm. The results of digital integration of the signals are shown in Fig. 5.5c. (Quite fortuitously, the average integration for the OH resonance and the standard iodoform come out to be the same. Normally this would not be the case).

| Signal (δ ppm) | Average Integration* |
|---|---|
| 5.4 $CHI_3$ (iodoform)** | 8.71 |
| 4.9 $CD_3OH$ | 8.71 |
| 3.5 $CH_2HOD$ | 12.00 |

* Average of seven separate integrations with a standard deviation of <0.05.
** 0.0499 g dissolved in 0.8898 g $CD_3OD$

**Fig. 5.5c.** *Integration results in $CH_3OD$ purity check*

To calculate the purity of the solvent we proceed as follows:

0.0499 g of $CHI_3$ was dissolved in 0.8898 g $CD_3OD$

ie $1.299 \times 10^{-4}$ mol $CHI_3$ ($M_r = 384$)

in $2.472 \times 10^{-2}$ mol $CD_3OD$ ($M_r = 36$)

∴ If $1.299 \times 10^{-4}$ mol of standard gives an integration of 8.71 there is $(8.71/8.71) \times 1.299 \times 10^{-4}$ mol of $CD_3OH = 1.299 \times 10^{-4}$

and $(12.00/8.71) \times 1.299 \times 10^{-4}$ mol of $CD_2HOD$

$= 1.789 \times 10^{-4}$

∴ percentage $CD_3OH$

$= (1.299 \times 10^{-4}/2.472 \times 10^{-2}) \times 100 = 0.53\%$

and percentage $CD_2HOD$

$= (1.789 \times 10^{-4}/2.472 \times 10^{-2}) \times 100 = 0.72\%$

∴ Total H content (mole %) in $CD_3OD$ sample

$= 0.53 + 0.72 = 1.25\%$

We can interpret this result as meaning that 1.25% of the methanol molecules contain one hydrogen atom.

So this sample of $CD_3OD$ is quite high in protons by present day standards where less than 0.2% is the norm. In fact this particular sample had been bought over ten years previously and 'lost' in a colleague's cupboard for that length of time.

In summary, proton nmr is an extremely useful and easy way of determining the extent of deuteration in a sample. The method has been used widely in mechanistic and biological studies.

**SAQ 5.5a**

The ideal internal standard for quantitative nmr spectroscopy probably does not exist. For each of the following give at least one disadvantage for its use as an internal standard

(*i*) TMS

(*ii*) Iodoform

(*iii*) Sodium ethanoate ($Na^+CH_3CO_2^-$)

(*iv*) Malonic acid ($CH_2(CO_2H)_2$)

**SAQ 5.5b**

The following results were obtained in a purity check on a sample of $CD_3OD$.

0.0601 g of $CHI_3$ dissolved in 0.8528 g $CD_3OD$ gave the following resonance intensities:

$CHI_3$, 10.3; $CD_3OH$, 8.64; $CD_2HOD$, 11.9.

Which of the choices listed below is the most likely proton content (mole %) of this sample?

(*i*) 1.31%

(*ii*) 0.55%

(*iii*) 0.76%

(*iv*) 1.25%

## 5.6. THE CASE OF THE FAR-SIGHTED FLUORINE

All our examples so far have been based on proton nmr. Now I want to discuss in some detail the analysis of another polymer system—polyether polyols—in which proton nmr fails, but fluorine-19 nmr offers an easy reliable analytical method for the determination of the percentage primary hydroxyl in the total hydroxyl content. This illustration will also show that the fluorine-19 nucleus is extremely sensitive to small structural changes even though the change might be many bonds away from the fluorine nucleus.

Polyether polyols are important industrial polymers which are made from two monomers, ethylene oxide and propylene oxide (Fig. 5.6a). If ethylene oxide is polymerised on its own polyethylene glycols (PEGs) are formed, and similarly propylene oxide given polypropylene glycols (PPGs). Most commercial polymers of this type are copolymers containing different proportions of PEG and PPG, and have different chain lengths (molecular masses).

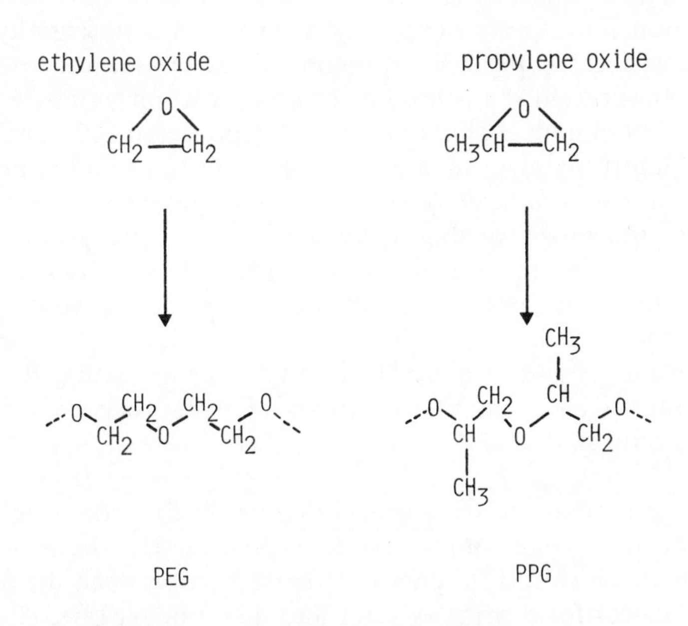


**Fig. 5.6a.** *Synthesis of PEG's and PPG's*

If ethylene oxide and propylene oxide are condensed together, or sequentially, then the polymer chain can end in two ways. (Fig. 5.6b). Either there will be a $-CH_2OH$, a primary hydroxyl, or a $-CH(CH_3)OH$, a secondary hydroxyl. As the subsequent reactivity of these polymers (especially in the manufacture of polyurethane foams) depends markedly on the relative amounts of primary and secondary hydroxyl there is a clear need for a convenient and accurate method of analysis for primary/secondary hydroxyl content.

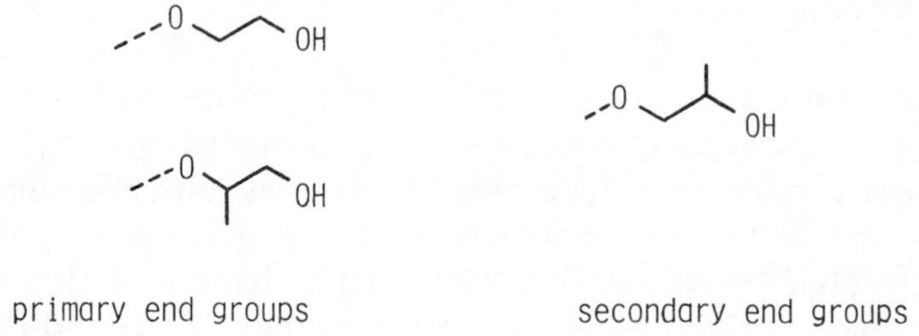

**Fig. 5.6b.** *Primary and secondary hydroxyls in polyether polyols*

Wet chemical analysis methods tend to be slow and inaccurate, but proton nmr spectroscopy was considered a potentially useful technique because protons on secondary carbon atoms resonate to slightly lower field than those on primary carbon atoms. For alcohols typical chemical shifts are $\delta = 3.4$ ppm and 3.5 ppm respectively. Unfortunately, the large number of protons on the polyether chain give rise to a large resonance at 3.4 ppm which overlaps and obscures the much smaller resonances of the end group protons (Fig. 5.6c). Spinning side bands and carbon-13 satellites also cause interference. Thus proton nmr cannot be used effectively.

Primary and secondary alcohols react very readily with trifluoracetic anhydride to give quantitative yields of trifluoroacetyl esters and trifluoracetic acid.

Now even though the structural difference is five bonds away from the fluorine nucleus and across an ester function there is a fairly large chemical shift difference of 0.3–0.4 ppm between the fluorine-19 resonances for a primary ester and a secondary one. Also there cannot be any interfering resonances from the rest of the polyether

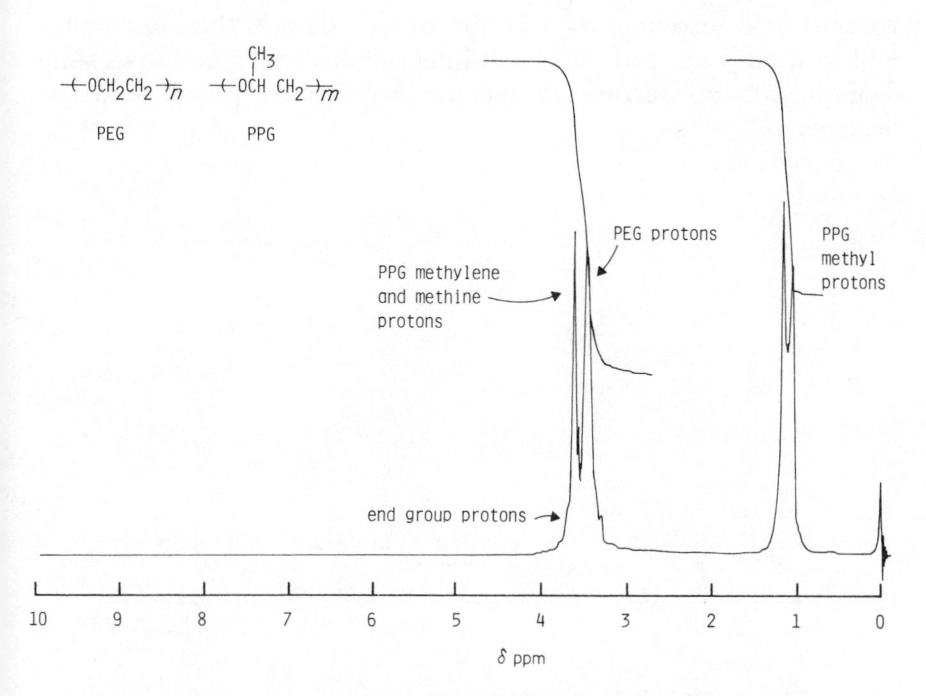


**Fig. 5.6c.** *Proton nmr spectrum of a polyether polyol*

$$\sim O\!-\!CH_2CH_2\!-\!OH \xrightarrow{CF_3C(O)OC(O)CF_3} \sim O\!-\!CH_2CH(H)\!-\!O\!-\!C(=O)\!-\!CF_2(F) + CF_3CO_2H$$

$$\sim O\!-\!CH_2CH(CH_3)\!-\!OH \xrightarrow{CF_3C(O)OC(O)CF_3} \sim O\!-\!CH_2CH(CH_3)\!-\!O\!-\!C(=O)\!-\!CF_2(F) + CF_3CO_2H$$

**Fig. 5.6d.** *Primary and secondary trifluoroacetyl esters*

chain as it contains only protons, no fluorines. Luckily, the trifluoroacetic acid resonates over 1 ppm to high field of the ester signals and so it does not interfere with integrations. Fig. 5.6e shows some examples of the fluorine-19 spectra of polyether polyol trifluoroacetates.

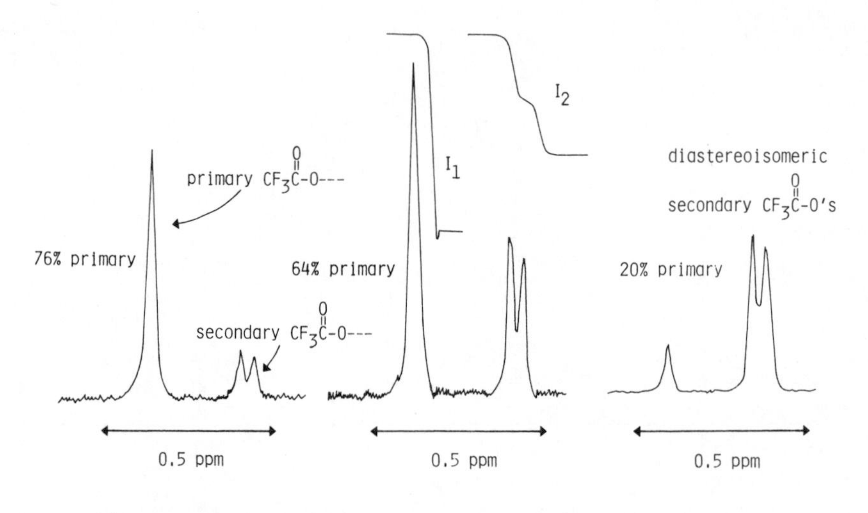


**Fig. 5.6e.** *Fluorine-19 nmr spectra of polyether polyol trifluoroacetates*

Before we look at some specific quantitative results there are two further interesting features in this example. Firstly, the formation of the trifluoroacetyl esters, as well as making the arithmetical analysis easy, has multiplied the number of spin-active nuclei involved in the analysis. What I mean here is that if we were looking at a primary $CH_2$ or secondary CH signal they would arise from two and one spin-active nucleus respectively. Now with the trifluoroacetyl esters, there are three nuclei representing both the primary and secondary centres. So we have, effectively increased the sensitivity of the experiment. It is interesting to speculate on other examples of this 'chemical amplification of sensitivity'. Could, for instance, a hexafluoro group be introduced from hexafluoroacetone?

Secondly, a closer look at the fluorine-19 spectrum (Fig. 5.6e) shows that while the low field peak is a singlet (and this can be shown to be the resonance for the primary trifluoroacetyl ester) the resonance for the secondary ester is double peak. The origin of this double peak is an extremely interesting stereochemical point and reflects again the sensitivity of fluorine-19 chemical shifts to structural differences. Fig. 5.6f shows that if a PPG fragment ends the polymer chain the next to last methyl group could have two stereochemical configurations in space with respect to the final methyl group, ie diastereoisomers are possible. This structural difference is picked up eight bonds away by the fluorine nucleus—a far-sighted effect indeed.

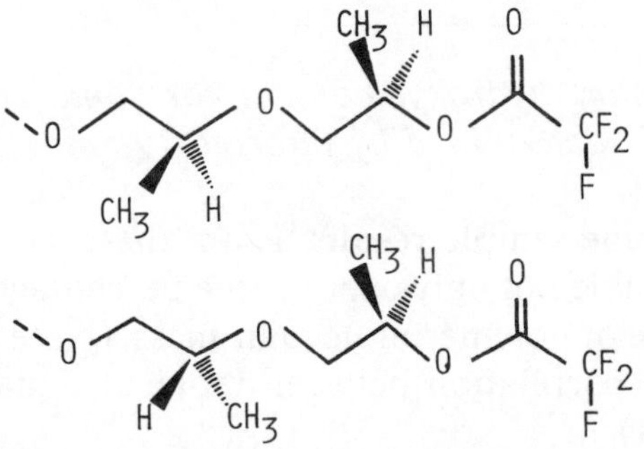

**Fig. 5.6f.** *Diastereoisomeric end groups in PPGs*

Nevertheless, the last point is not a problem as far as integration is concerned, the primary and second resonances are well separated so there is easy integration and the calculation is obvious.

As we wish to determine primary hydroxyl as a percentage of the total hydroxyl content we take the ratio of the integration of the trifluoroacetyl peak derived from the primary hydroxyl to the total trifluoroacetyl integration and multiply by 100, as shown in the following equation.

$$\% \text{ primary} = \frac{I_1}{I_1 + I_2} \times 100\%$$

where $I_1$ and $I_2$ are the integrations for the primary and secondary resonances respectively.

| Molecular Mass | % Primary |
|---|---|
| 300 | 8 |
| 2000 | 58 |
| 3100 | 20 |
| 4000 | 3 |
| 4800 | 75 |
| 6000 | 70 |

**Fig. 5.6g.** *Primary hydroxyl content for some polyether polyols determined by fluorine-19 nmr*

Fig. 5.6g gives some sample results. From these we can see that the method is applicable not only over a large percentage range, but also over a wide range of polymer molecular mass. (Note that there is not meant to be any correlation between molecular mass and primary hydroxyl content).

This example has brought two main analytical themes of this Unit, quantitative and qualitative applications, together. While the quantitative results are easily derived, as they have been in most of these cases, the qualitative aspects which underpin the quantitative method rely on quite subtle structural differences—differences which nmr spectroscopy, in this case using fluorine-19, can easily detect.

**SAQ 5.6a**

Fill in the gaps in the following description from the list below.

The determination of primary and secondary ... (1) ... of polyether polyols by fluorine-19

⟶

**SAQ 5.6a (cont.)**

nmr spectroscopy depends on the ... (2) ... in the trifluoroacetyl esters of the polyether polyols sensing structural differences ... (3) ... distant. Chemical shift differences of ... (4) ... result between the primary and secondary esters, and these signals can be easily ... (5) ... so allowing quantitative determinations.

(*i*) four bonds

(*ii*) five bonds

(*iii*) 3–4 ppm

(*iv*) integrated

(*v*) hydroxyl content

(*vi*) 0.3–0.4 ppm

(*vii*) fluorine nuclei

**SAQ 5.6b**

A sample of a polyether polyol treated with trifluoroacetic anhydride gave the following set of digital integration values for the primary and secondary trifluoroacetyl esters. What is the percentage of primary hydroxyl?

Average of 5 integrations for primary resonance = 12.4

Average of 5 integrations for secondary resonance = 108.2

(*i*) 10.3%

(*ii*) 11.5%

(*iii*) 89.7%

(*iv*) 23.2%

## 5.7. SUMMARY OF PART 5

We have covered a good deal of material in this Part of the Unit, using drugs, polymers, a measure of whisky and a bottle of claret amongst others to illustrate the diversity of quantitative nmr and the general simplicity of deriving quantitative data.

The main considerations that we have used in these quantitative analytical problems can be summarised as follows:

(*a*) The area of an nmr peak is proportional to the number of spin-active nuclei giving rise to it, and so to the molar concentrations of the spin-active nuclei. (Although this does not hold for routine carbon-13 spectra).

(*b*) The areas of peaks as measured by integrations can be used to measure similar concentrations reasonably accurately.

(*c*) The arithmetical/algebraic procedures for dealing with quantitative determinations is usually easily worked out from first principles for individual cases.

(*d*) The determination of absolute concentrations of components in a mixture requires the addition of a known mass of an internal reference, while relative concentrations of components do not require an internal reference.

**Objectives of Part 5**

Now that you have completed Part 5 you should be able to:

- appreciate that the integration of nmr peak areas is an important source of quantitative data;
- appreciate the range of problems that can be tackled by quantitative nmr;

- describe several examples of quantitative nmr, viz:

  - the determination of the relative amounts of phenacetin and aspirin in an APC tablet;

  - the determination of ethanol/water mixtures;

  - the determination, via the method of standard additions, of stearic acid in a stearic acid/polyethylene mixture;

  - the determination of the molecular mass of a polymer;

  - the determination of the extent of deuteration of a solvent, via the use of an internal standard;

  - the determination of primary hydroxyl content in samples of polyetherpolyols via fluorine-19 nmr;

- apply the principles of quantitative nmr to problems similar to those in the above objectives.

# Self Assessment Questions and Responses

**SAQ 1.1a**

Which of the following statements about the nmr phenomenon are correct?

Circle T (true) or F (false).

(*i*) Nuclear magnetic resonance is to do with the absorption of radiowaves by certain kinds of nuclei when they are in a strong magnetic field.

T / F

(*ii*) The precise frequency that a nucleus will absorb, when placed in a magnetic field is given by the equation $\nu = \dfrac{\gamma}{2\pi} B_0$

T / F

(*iii*) All nuclei can undergo nmr.

T / F

(*iv*) The number of spin states for a nucleus is given by the expression $(2I + 1)$ where $I$ is the spin quantum number of the nucleus.

T / F

⟶

**Response**

Statement (*i*) is true; it is more or less the first sentence of this section, and could be taken as a rough definition of the phenomenon of nmr. Note that it does leave a number of points unexplained: what kinds of nuclei? Why is energy absorbed? What is a strong magnetic field? The answers to these points are the subject of the first part of this section.

Statement (*ii*) is also correct, given that the nucleus in question is spin-active. This equation predicts the fundamental resonance frequency, or Larmor frequency, of a nucleus given the strength of the magnetic field ($B_O$) and the magnetogyric ratio ($\gamma$) of the nucleus.

Statement (*iii*) is incorrect, not all nuclei are capable of nuclear magnetic resonance because some elemental nuclei have a spin quantum number of zero (eg, carbon-12) and this means that they can have only one spin state. If a nucleus has a non-zero spin quantum number then it can be spin-active and undergo resonance. The number of spin states being given by the expression $(2I + 1)$.

So statement (*iv*) is correct.

***********************************

**SAQ 1.1b**

Boron-11 has a spin quantum number of $\frac{3}{2}$. How many spin states can it have?

(*i*) 6

(*ii*) 5

(*iii*) 4

(*iv*) 3 ⟶

**Response**

This should have been easy, for all you had to do was apply the ($2I$ + 1) expression. If $I$ is $\frac{3}{2}$ then the expression reduces to (3 + 1), 4. So answer (*iii*) was correct. If you did not have that, then you probably made a simple arithmetical error. As a matter of interest, boron has another isotope, boron-12, which has a spin quantum number of 3. Applying the expression ($2I$ + 1) gives boron-12 seven spin states—a much more complex state of affairs than protons and carbon-13.

***********************************

**SAQ 1.1c**

From the list of words/symbols fill in the spaces (1) to (4) in the following diagram which is an energy description of the nmr phenomenon. Note that there are some items in the list that *do not* fit at all.

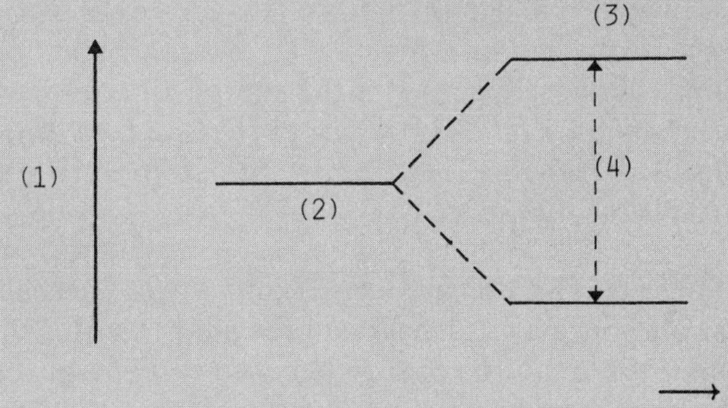

(1)
(2)
(3)
(4)
→

**SAQ 1.1c (cont.)**

(*i*) $\Delta E \propto \Delta B_O$

(*ii*) Energy

(*iii*) magnetic field

(*iv*) $h\nu \propto B_O$

(*v*) zero field

(*vi*) $\nu \propto \gamma/2\pi$

**Response**

Your answer should have been based on Fig. 1.1b. The space (1) is the energy axis so (*ii*) is correct here. Space (2) represents the system with no magnetic field so (*v*), zero field is correct. Space (3) represents the system when placed in a magnetic field so (*iii*) is the required answer. Space (4) is the energy difference between the spin states. Now in Fig. 1.1b this is given as $\Delta E \propto B_O$, but in the text we applied Planck's Law, $\Delta E = h\nu$, so $h\nu \propto B_O$ and answer (*iv*) is correct for this space.

Response (*i*) and (*vi*) are incorrect in any of the spaces. If you chose (*i*) at all you may not have appreciated that the simple $\Delta$ usually means a difference or change between two energy levels (like $\Delta G^+$ in thermodynamics). So $\Delta B_O$ would imply a change in the magnetic field.

If you chose (*vi*) then you may not have appreciated that $\gamma/2\alpha$ is a constant for a given nucleus and a frequency cannot be proportional to a constant.

***********************************

**SAQ 1.1d**

Given the Boltzmann constant to be

$1.3806 \times 10^{-23}$ J K$^{-1}$

and Planck's constant to be

$6.6262 \times 10^{-34}$ Js

which of the following is the population difference between the lower and upper spin states for spin-active nuclei resonating at 100 MHz and 33 °C in a 2.348 Tesla field?

(*i*) about 7 per million

(*ii*) about 9 per million

(*iii*) about 16 per million

(*iv*) about 145 per million

**Response**

Answer (*iii*) is correct.

We have to do the calculation using the Boltzmann equation:

$$N_2/N_1 = 1 - \Delta E/kT$$

Now for $\Delta E$ we can use $h\nu$ ie

$h \times 100 \times 10^6$ because we are told the nuclei are resonating at 100 MHz. So the equation becomes:

$$N_2/N_1 = 1 - \frac{6.6262 \times 10^{-34} \times 100 \times 10^6}{1.3806 \times 10^{-23} \times 306}$$

Remember that $T$ must be the absolute temperature. Evaluating the above expression gives :

$$N_2/N_1 = 1 - 1.57 \times 10^{-5}$$

or $$N_1 - N_2 = 1.57 \times 10^{-5} \times N_1$$

so per million of $N_1$, $N_1 - N_2 = 1.57 \times 10^{-5} \times 10^6$

$$= 15.7, \text{ ie about 16—Answer } (iii)$$

Answer (*i*) is the population difference if $\nu$ = 40 MHz, answer (*ii*) is the difference at 60 MHz. Answer (*iv*) is what you might have calculated if you forgot to use the absolute temperature and had put $T$ = 33.

*********************************

**SAQ 1.3a**

Indicate by circling T for true or F for false which of the following statements concerning chemical shift are correct.

(*i*) The chemical shift of a nucleus is defined as the shift in ppm from the resonance frequency of a standard, usually TMS.

T / F

(*ii*) The chemical shift of TMS is always 0 ppm.

T / F

(*iii*) The chemical shift of a nucleus will always be to the left of the reference in an nmr spectrum.

T / F

⟶

**SAQ 1.3a (cont.)**

(*iv*) When looking at an nmr spectrum the resonance frequency increases on going from left to right.

T / F

**Response**

Statement (*i*) is true. Eq. 1.3a in the text shows you that the chemical shift was defined as being equal to

$$\frac{\nu_R - \nu_{TMS}}{\nu_{spectrometer}} \times 10^6$$

This expression is a dimensionless quantity and is given the units $10^{-6}$, ie ppm.

Statement (*ii*) is true. By definition in the above expression if $\nu_R = \nu_{TMS}$ then $\nu_R - \nu_{TMS} = 0$. So the whole expression is 0.

Statement (*iii*) is false. While the great majority of proton and carbon-13 resonances do occur to the left (downfield) of TMS some structures are more shielded than TMS and so will resonate to higher field, ie, to the right of TMS. Some cyclopropyl groups and organometallic compounds fall into this type. They would have negative $\delta$ values.

Statement (*iv*) is false. Fig. 1.3b shows that for a normal nmr spectrum frequency increases to the left.

***********************************

**SAQ 1.3b**

The proton in chloroform ($CHCl_3$) is found to resonate at $\delta$ = 7.25 ppm on a 60 MHz instrument.

Calculate:

(*i*) The frequency difference in Hz between the $CHCl_3$ resonance and that of TMS.

(*ii*) The frequency difference between the $CHCl_3$ and TMS resonances if the operating frequency were (*a*) 100 MHz, (*b*) 220 MHz.

(*iii*) The chemical shift, $\delta$, of the $CHCl_3$ resonance at (*a*) 100 MHz, (*b*) 220 MHz.

**Response**

(*i*) At 60 MHz 1 ppm = 60 Hz

a chemical shift of 7.25 ppm = 60 × 7.25 Hz

= 435.0 Hz

(*ii*) To answer this part we need to remember that resonance frequencies are proportional to magnetic field strength and so proportional to operating frequencies.

$$\text{so} \quad \frac{\text{frequency difference at 100 MHz}}{\text{frequency difference at 60 MHz}} = \frac{100}{60}$$

$\therefore$ frequency difference at 100 MHz =

$$\frac{100 \times 435}{60} = 725 \text{ Hz}$$

Similarly, frequency difference at 220 MHz =

$$\frac{220 \times 435}{60} = 1595 \text{ Hz}$$

(*iii*) By definition at 100 MHz the chemical shift is

$$\frac{(725 - 0) \times 10^6}{100 \times 10^6} = 7.25 \text{ ppm}$$

Similarly, at 220 MHz $\delta$ = 7.25 ppm

Note that when chemical shifts are defined as in the text they are effectively constant at any operating frequency.

***********************************

**SAQ 1.3c**

A forgetful nmr operator omitted TMS from a sample for nmr analysis. The sample was dissolved in $CDCl_3$ and the operator correctly positioned the resonance of residual $CHCl_3$ at $\delta$ = 7.25 ppm. Two further peaks were observed in the sample, one 2 ppm downfield from the $CHCl_3$ resonance and one 3.5 ppm upfield from it. What were the chemical shifts of these two resonances?

**Response**

This question should have had you thinking about the terms *upfield* and *downfield.* Again reference to Fig. 1.3b is in order here. Resonances are said to be downfield if they are to the left while upfield resonances are to the right. So the resonance at 2 ppm downfield from the $CHCl_3$ resonance is $\delta$ = 9.25 ppm while the 3.5 ppm upfield resonance is $\delta$ = 3.75 ppm.

***********************************

**SAQ 1.4a**

Rank the spin-active nuclei in the following sets in their likely chemical shift order lowest to highest (ie most shielded to least). Note that precise chemical shifts are not required.

eg $-CH_2-$, $-CH_3$, $-CH$

Answer $-CH_3$ > $-CH_2-$ > $-CH$
for both proton and carbon-13.

(*i*) benzene ($C_6H_6$), methanal ($CH_2O$), ethyne ($HC{\equiv}CH$)

(*ii*) $CHCl_3$, $CH_2Cl_3$, $CCl_4$, $CH_3Cl$

(*iii*) Methanal ($CH_2O$), acetone ($CH_3COCH_3$), methyl formate ($CH_3OCOH$).

(*iv*) the different protons only in salicylic acid $C_6H_4(OH)(CO_2H)$ (treat the aromatic protons as being just one type):

OH
$CO_2H$

**Response**

The answers to this question are all to do with the various factors affecting chemical shifts. Thus in (*i*) you should have noted that the three compounds all have different hybridisations and reference to Fig. 1.4a would have shown you that for both the protons and the carbons the order of chemical shifts would be:

$$HC{\equiv}CH > C_6H_6 > CH_2O$$

For set (*ii*) you should have seen that the number of electronegative substituents on the carbon atom is increasing, so the polarity of the carbon and associated hydrogens becomes increasingly positive. The order for both proton and carbon chemical shifts must be:

$$CH_3Cl > CH_2Cl_2 > CHCl_3 > CCl_4$$

The last chemical, of course, has no protons and so will not show a proton nmr, but it does have the most deshielded carbon atom and so the highest carbon-13 chemical shift.

There are a few more types of protons and carbons in set (*iii*), but taking them one by one we can sort them out. First the protons; according to the hybridisation factor the methanal and formate protons will be deshielded with respect to the methyl protons in acetone and methyl formate. So $CH_3$ (acetone, formate) > CH (methanal, formate).

Within these two groups of two the methyl group of the formate ester will be deshielded with respect to the acetone protons because of the electronegativity of the oxygen. The CH of the formate will be shielded with respect to the aldehyde protons because of a mesomeric effect.

So, for the protons:

$CH_3$(acetone) > $CH_3$(formate) > CH(formate) > CH(methanal)

is the correct order from most shielded to least.

For the carbon atoms, of which there are five different types, we can apply the same arguments. On hybridisation grounds we would expect:

$CH_3$ (acetone, formate) > C=O (acetone, formate, methanal)

On inductive effect grounds the methyl carbons of acetone would be more shielded than that of the methyl carbon of the formate ester. Similarly, on mesomeric ground the formate carbonyl carbon should be more shielded than the ketone or aldehyde carbonyl. These latter carbon atoms might be expected to be very similar in chemical shift and in fact our ability (from these notes) to predict an order runs out at this point. So for the carbons:

$CH_3$(acetone) > $CH_3$(formate) > C=O(formate) > C=O(aldehyde, ketone).

If you managed all of set (*iii*) absolutely correctly, well done indeed.

Set (*iv*) is included to bring in some hydrogen-bonded protons and reference to Figs. 1.4a and 1.4e should have given you the order for the proton chemical shifts as:

—OH > aromatic CH > $—CO_2H$

If you had that order consider yourself correct. However, at some concentrations the phenolic and carboxylic resonances combine to form a very broad band underneath the aromatic resonance!! That is the trouble with labile protons—you can never be absolutely sure of their chemical shifts.

**********************************

**SAQ 1.4b**

Which of the following statements about anisotropy in chemical bonds are correct? Circle T (true) or F (false).

(*i*) Anisotropy in chemical bonds means the electron density and hence the shielding effect of bonds is different in different regions around the bond.

T / F

(*ii*) Anisotropy always leads to spin-active nuclei being deshielded.

T / F

→

**SAQ 1.4b (cont.)**

(*iii*) Aldehyde protons and carbonyl carbon atoms owe their chemical shifts solely to anisotropic effects

T / F

(*iv*) Anisotropy occurs only in $\pi$-bonds.

T / F

**Response**

Statement (*i*) is true. It provides us with a working explanation of the term anisotropy when applied to nmr matters. The word isotropic means having the same physical properties in all directions. Hence anisotropic means having different or varying properties in different directions. In connection with chemical shifts the property in question is the electron density associated with chemical bonds, hence differences in shielding.

Statement (*ii*) is false. Whether a spin-active nucleus is shielded or deshielded by anisotropic effects depends on the anisotropy. Reference to Fig. 1.4b shows you that, for a $\pi$-bond, if a spin-active nucleus were to be held above the plane of the $\pi$-bond then it would be shielded. Nuclei in the plane of the $\pi$-bond would be deshielded.

Statement (*iii*) is false. While the chemical shifts are largely caused by the $\pi$-bond's anisotropy, the electronegativity of the oxygen atom also reduces electron density at the carbonyl carbon atom and so enhances the anisotopic effect.

Statement (*iv*) is also false. Our discussion has centred around $\pi$-bonds, but every bond has its own anisotropy, sometimes resulting in pronounced chemical shift variations as with $\pi$-bonds, and sometimes in much smaller changes. Interested readers are referred to the recommended textbooks for further details.

***********************************

**SAQ 1.5a**

Which of the following statements about spin–spin coupling are correct?

Indicate your answer by circling T (true) or F (false).

(*i*) Spin–spin coupling arises because one spin-active nucleus interacts with the spin states of nearby spin-active nuclei.

T / F

(*ii*) Spin–spin coupling depends on the size of the external magnetic field $B_O$.

T / F

(*iii*) The coupling constant, $J$, is always measured in Hertz.

T / F

**Response**

Statement (*i*) is true. The magnetic field experienced by a spin-active mucleus is going to be modified by the magnetic fields of neighbouring nuclei. The actual magnetic field of these adjacent nuclei depends on their spin states.

Statement (*ii*) is false. If you think carefully about spin coupling you should see that it exists between nuclei all the time, whether or not there is an external magnetic field. The nuclear property of spin, and so magnetic moments, is a fundamental nuclear property. The nmr experiment allows us to observe this coupling.

Statement (*iii*) is true, but you should be aware of why coupling constants are measured in Hertz, and not ppm. As we have seen chemical shifts when measured in frequency units are dependant on the operating frequency, and magnetic field strength and so we measure them as a ratio of observed frequency to operating frequency.

Coupling constants are independent of the operating frequency and magnetic field strength and can be measured in the more fundamental frequency units.

*********************************

**SAQ 1.5b**

The spectrum of a sample sent for nmr analysis showed two peaks separated by 0.1 ppm at 60 MHz. If the two peaks are in fact a doublet arising from coupling which *one* of the following would be the splitting at 220 MHz?

(*i*) 0.1 ppm

(*ii*) 6 Hz

(*iii*) 22 Hz

**Response**

Remember that coupling constants do not depend on operating frequency, so if the splitting of 0.1 ppm is a coupling constant then as 0.1 ppm at 60 MHz is 6 Hz this will be the coupling constant at any frequency. So answer (*ii*) is correct. Answers (*i*) and (*iii*) are related as 0.1 ppm at 220 MHz is 22 Hz. They would have been the splitting at 220 MHz had the doublet been due to two different chemical shifts.

*********************************

**SAQ 1.5c**

In the compound dipropyl ether

$$(CH_3CH_2CH_2OCH_2CH_2CH_3)$$

the central methylene protons' resonance is a sextet. Which of the following is the likely intensity ratio for the six peaks?

(*i*) 1:2:3:3:2:1

(*ii*) 1:2:1:1:2:1

(*iii*) 1:7:9:9:7:1

(*iv*) 1:5:10:10:5:1

**Response**

Answer (*iv*) is correct.

This question was just to test your arithmetical prowess when using Pascal's Triangle. You have to calculate the sixth level of coefficients. Referring to Fig. 1.5f start the sixth row with 1, add together the two coefficients immediately above to give 5, repeat for the next coefficient to give 10 and so on, finishing with a 1. So answer (*iv*) is correct. If you chose any of the others you will need to brush up on your arithmetic.

***********************************

**SAQ 1.6a**

Concerning the molecular fragment

$$CH_3CH_2CH_2—,$$

which of the following statements about spin coupling are correct. Indicate your answer by circling T (true) or F (false).

(*i*) The proton spectrum will show a quartet of peaks for the methyl resonance.

T / F

(*ii*) The off-resonance carbon-13 spectrum will show a quartet of peaks for the methyl resonance.

T / F

(*iii*) The proton spectrum will probably show a sextet of peaks for one of the methylene resonances.

T / F

(*iv*) The off-resonance carbon-13 spectrum for the methylene carbons will show two triplet resonances with ratios of intensities 1:2:1.

T / F

**Response**

The methyl group protons are coupled to the two protons of the adjacent carbon atom. Applying the $(n + 1)$ rule, where $n = 2$ gives three peaks for the methyl resonance not four. Thus statment (*i*) is false. Again using the $(n + 1)$ rule to discover the number of lines for the carbon-13 methyl resonance we have $n = 3$ so $n + 1 = 4$, ie a quartet should be observed. So statement (*ii*) is true.

Statement (*iii*) is true provided we recognise the word 'probably'. The protons on the central carbon atom have five adjacent protons and applying the ($n$ + 1) rule allows us to predict six peaks in the splitting pattern. However, we are making an assumption that the coupling constant between the methyl protons and those on the central carbon atom is the same as that between the end methylene protons and those on the central carbon atom. If the coupling constant were different how would this affect the pattern? The answer is not too difficult to see. The methyl protons would split the methylene resonance into a quartet and the end methylene protons would split each of the quartets' lines into a triplet. So a total of twelve lines would result. (Fig. 1.6b). Depending on the relative sizes of the two couplings a quartet of triplets, or a triplet of quartets, or some overlapping pattern might result.

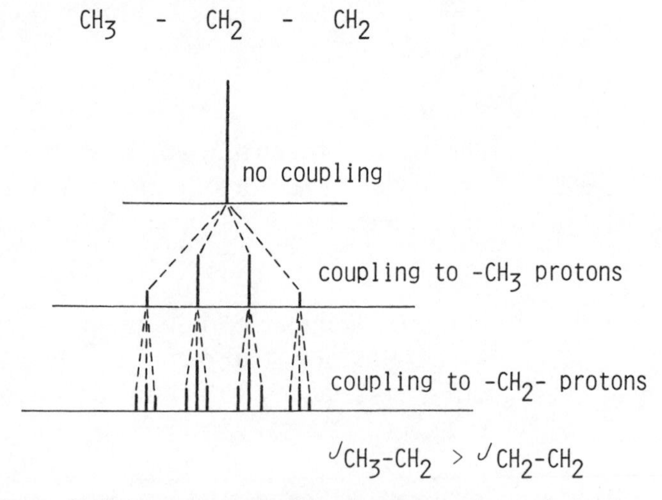


**Fig. 1.6b.** *Splitting pattern for the central methylene protons of* $CH_3CH_2CH_2-$

*NB*—Intensities not to scale.

Statement (*iv*) is unlikely to be true. While we would expect two triplets in the off-resonance carbon-13 spectrum the peaks in the triplets are unlikely to be in a 1 : 2 : 1 ratio as predicted by Pascal's Triangle. Various instrumental factors cause distortion of relative intensities.

*************************************

**SAQ 1.7a**

Which of the following will be the correct number of lines observed for the methyl carbon resonance in the carbon-13 spectrum of hexadeuteroacetone ($CD_3COCD_3$)?

(*i*) 4

(*ii*) 6

(*iii*) 7

(*iv*) 13

**Response**

Answer (*iii*) is correct.

We cannot use the $(n + 1)$ rule here because the spin quantum number of deuterium is 1, not $\frac{1}{2}$. So we have to use the more general version $(2nI + 1)$. In hexadeuteroacetone the two deuteromethyl groups are symmetrically equivalent so we need only calculate $2nI + 1$ for one group. ($n = 3$, $I = 1$). So there should be seven lines—answer (*iii*) is correct. If you chose (*i*) then you probably forgot the deuterium has a spin of 1; if you chose (*ii*) you might have thought that the number of lines would be the same as the total number of coupling nuclei. That is incorrect. If you chose (*iv*) then you probably thought that as there are six equivalent deuterons '$n$' should equal six in the formula. So $2nI + 1$ would equal 13. Now this interpretation is wrong because the $(2nI + 1)$ coupling rule states that '$n$' must be the number of equivalent nuclei coupling to the nucleus under consideration. It is impossible to have six equivalent deuterons coupling to one carbon-13 nucleus.

**********************************

**SAQ 1.8a**

Fill in the gaps in the following statement about relaxation phenomena from the list of phrases/symbols supplied. Note that some of the list are quite inappropriate responses.

Two relaxation processes for nuclei are ... 1 ... and ... 2 ... relaxation. In the first process ... 3 ... to the rest of the system while the second process involves ... 4 .... The processes are characterised by relaxation times ... 5 ... and ... 6 ....

(*i*) energy is lost by

(*ii*) energy is lost to

(*iii*) spin–lattice

(*iv*) $T_1$

(*v*) spin inversion

(*vi*) $T_{\frac{1}{2}}$

(*vii*) spin–spin

(*viii*) $T_2$

(*ix*) exchange of spins

**Response**

Space 1 should be phrase (*iii*), spin–lattice. Space 2 should be phrase (*vii*), spin–spin. Space 3 should be phrase (*ii*), energy is lost to. Note that phrase (*i*), energy is lost by, is incorrect in this context as the excess energy of the high spin state gets dissipated into the rest of the molecular system. Space 4 should be phrase (*i*), exchange of spins. Note that phrase (*v*), spin inversion, might just be allowed as meaning the same as exchange of spins, but really refers to an individual nucleus undergoing a change of spin.

Space 5 and space 6 should be (*iv*) and (*viii*) respectively, $T_1$ and $T_2$. The option (*vi*), $T_{\frac{1}{2}}$, might refer to a half-life in some other processes, but is not used in the context of relaxation times.

*********************************

**SAQ 1.8b**

Which of the following statements about relaxation phenomena are true?

Indicate your answer by circling T (true) or F (false).

(*i*) The relaxation time for a spin-active nucleus is the time taken for the nucleus to change from a high energy to low energy spin state.

T / F

(*ii*) The relaxation time for a spin-active nucleus depends on the type of nucleus.

T / F

(*iii*) Spin–spin relaxation is slower than spin–lattice relaxation.

T / F

(*iv*) Spin–lattice relaxation times depend on the viscosity of the sample.

T / F

⟶

**Response**

Statement (*i*) sounds tantalisingly correct, but, unfortunately is false. Reference to Eq. 1.5 shows that relaxation times are defined as a parameter in a rate equation, rather in the way that radioactive isotopes have half-lives. The high energy spin-active nuclei in a large sample are going to decay over a period of time, some very quickly, some much more slowly; the relaxation time gives a measure of that exponential decay. Beware also of another false interpretation of statement (*i*). For a spin-active nucleus the resonance phenomenon from low to high spin state is essentially instantaneous. The relaxation time is a measure of the time spent in the high energy state, but not of the time taken in actually going from one state to the other.

Statement (*ii*) is true, different nuclei either isotopically different (fluorine-19, proton, etc) or structurally different (methyl protons, methylene protons etc) can have different relaxation times. We have not spent much time on this aspect because the measurement and interpretation of relaxation times is really beyond this text, but you should be aware that such information can be very valuable.

I hope you've not guessed incorrectly about statement (*iii*). It is one of those points where you know the answer is one or the other, but you can never remember which. In fact the statement must be false. First, the word 'slower' is not very good in the context, clearly 'longer' or 'greater than' is implied. Theoretically, spin–spin relaxation times cannot be longer than spin–lattice ones (for the same nucleus). A simple way of looking at it is to realise that spin–spin relaxation does not involve any energy change in the system, only spins are exchanged, so eventually some form of spin–lattice relaxation must take place to dissipate the extra energy.

If you correctly assessed the truth of statement (*iv*) then you are doing very well. Spin–lattice relaxation is intimately connected with the viscosity of the sample, but the theory of the relationship is really complex. It turns out that longer relaxation times occur for low viscosity solutions. The higher the viscosity a shorter relaxation time causes lines broadening—a well known effect in the spectra of neat liquids.

***********************************

**SAQ 1.8c**

A peak width at half height in an nmr resonance is found to be 0.3 Hz. Which of the following values is an approximate relaxation time for the resonance?

(*i*) 0.3 s

(*ii*) 3 s

(*iii*) 33 s

(*iv*) 0.03 s

**Response**

Answer (*ii*) is correct.

This should have been very straightforward because all you need do is apply Eq. 1.6 which says that approximate relaxation times and peak widths at half-height are inversely proportional.

Thus

$T_2$ = 1/peak width at half height = 1/0.3 = 3.33 s = 3 s

So answer (*ii*) is correct. If you chose any of the other answers you probably made an elementary arithmetical error.

***********************************

**SAQ 1.9a**

Which of the following statements concerning integration of nmr signals is correct?

Circle T (true) or F (false).

(*i*) For a proton nmr spectrum the areas under each resonance are proportional to the number of protons causing the signal.

T / F

(*ii*) Peak heights can be used instead of peak areas.

T / F

(*iii*) If a spectrum contains only one resonance there is little point in measuring its integration.

T / F

**Response**

Statement (*i*) is true and provides a concise definition of what integration in nmr spectroscopy is all about. In more general terms we could use the word spin-active rather than proton, bearing in mind that carbon-13 signals do not usually give reliable integrations

Alas, statement (*ii*) is false. We cannot use peak heights because they are not proportional to the number of spins causing the resonance. You might have come across the same idea in gas liquid chromatography, where it is the areas of the peaks in chromatograms which supply information about amount. In nmr spectroscopy (and GLC) it is possible to have two peaks the same height but different in area if the peak widths are different.

Statement (*iii*) is correct for if there is only one peak in a spectrum then we have no other with which to compare it. Absolute measurements of intensity are difficult to obtain in nmr spectroscopy and even in quantitative applications (Part 5) integration ratios are usually preferred.

************************************

**SAQ 1.9b**

The diagram below shows a proton nmr spectrum complete with integration curve. What is the likely ratio of protons?

(*i*) 5:2:3

(*ii*) 8:4:6

(*iii*) 2:1:1.5

(*iv*) 4:2:3

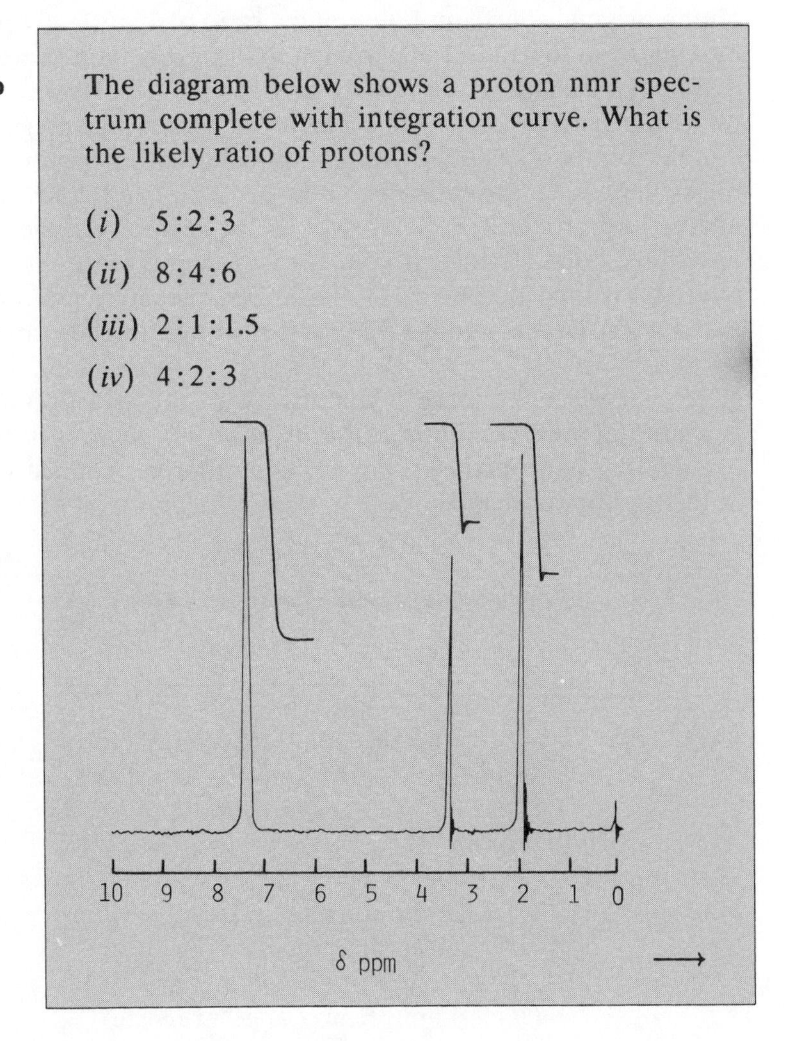

**Response**

Answer (*iv*) is correct.

The first thing you should have done was to have measured the heights of each step. It does not matter what units of length are used because a ratio is dimensionless. Using millimetres for convenience on the original spectrum I obtained a ratio of 42 : 20 : 29 for the steps moving from low field (left) to high field (right). To proceed we now divide by the smallest number to give 2.1 : 1 : 1.45. However, for simple compounds protons are going to be present in integer amounts (whole numbers) there is not going to be half a proton lurking in the structure. So the ratio becomes 4.2 : 2 : 2.9 and converting to the nearest integers 4 : 2 : 3. So answer (*iv*) is correct. Answer (*i*) would only have been attained if your measurements were incorrect. Answer (*ii*) is incorrect because division by two gives an integer ratio, and answer (*iii*) is wrong because it is not an integer ratio.

Note that we make these deductions with the assumption that this is a simple, pure compound. *Ratios need not be integer ratios if we are dealing with mixtures.* Part 5, Quantitative Applications, looks at this in greater detail.

*********************************

**SAQ 1.10a**

Which of the four nmr parameters, chemical shifts, spin–spin coupling constants, relaxation times and integrations could give you information on:

(*i*) the number of spin-active nuclei next to a resonating nucleus;

(*ii*) chemical exchange processes; ⟶

**SAQ 1.10a (cont.)**

(*iii*) functional groups;

(*iv*) the relative amounts of the components of a mixture.

**Response**

The number of spin-active nuclei next to a resonating nucleus, statement (*i*), is going to influence the number of adjacent spin states that the nucleus 'sees' and so the resonance will be split into fine structure by spin–spin coupling (Section 1.5). For chemical exchange processes, statement (*ii*), such as proton transfers then the relaxation times of the protons will vary and lines may be broadened (Section 1.8). Functional groups, statement (*iii*), in organic structures (and other types of material) will have protons, carbon-13, etc, in particular chemical environments and this will be reflected in their chemical shifts (Section 1.3). The relative amounts of the components of a mixture, statement (*iv*), may be obtained by the use of integration (Section 1.9) and is studied in greater depth in Part 5.

***********************************

**SAQ 2.1a**

A compound submitted for nmr analysis proved very difficult to dissolve in any one solvent. Eventually a pair of solvents was used and the spectrum had residual solvent resonances at $\delta$ = 2.1 ppm and 4.6 ppm. What was the most likely combination of solvents?

(*i*) $CDCl_3/CD_3OD$

(*ii*) acetone-$d_6$/benzene-$d_6$

(*iii*) acetic acid-$d_6$/$D_2O$

(*iv*) acetone-$d_6$/$D_2O$

**Response**

Using Fig. 2.1a to check the positions of residual resonances you should have found that only acetone-$d_6$ or acetic acid-$d_4$ could have given a resonance at about 2.1 ppm, while $D_2O$ is the most likely source of a trace resonance at 4.6 ppm. Thus either (*iii*) or (*iv*) might be correct, but the acetic acid-$d_4$ would be expected to give a further peak (to low field) and this is not observed. Thus (*iv*) is the correct answer.

**********************************

**SAQ 2.1b**

Consider the two structures below (both compounds are used extensively in the polymer industry) and answer the following questions as either true (T) or false (F). ⟶

**SAQ 2.1b (cont.)**

$$\text{X: } C_6H_4(CO_2H)_2 \qquad \text{Y: } HO_2C(CH_2)_6CO_2H$$

X

Y

(*i*) the % hydrogen in X is 8.05.

T / F

(*ii*) Sample Y will have to be examined at higher concentration than sample X to give comparable signal intensities in their proton nmr spectra.

T / F

(*iii*) For comparable carbon-13 nmr spectra solutions of the same molarity should be used.

T / F

(*iv*) The presence of acidic protons could be shown by treating $CDCl_3$ solutions with $D_2O$.

T / F

**Response**

Part (*i*) is false. In fact the figure given is the % hydrogen for compound Y. The correct figure is 3.6%.

Part (*ii*) is false. You need to know the % hydrogen concentration of X and Y. As we have seen Y has a greater % of hydrogen so that you would need more of X, not Y.

Part (*iii*) is true. You must consider the % carbon and hence % carbon-13 present in compounds X and Y. As the molecular masses are similar and both X and Y contain the same number of carbon atoms then similar concentrations should give comparable spectra.

Part (*iv*) is true as well. Acidic protons are readily exchanged and a broadish peak at about 12 ppm will disappear upon a $D_2O$ shake and be replaced by one at about 4.6 ppm (the HOD peak).

***********************************

**SAQ 2.1c**

Choose the correct word from the list below to answer the following questions:

(*i*) Solids and viscous liquids do not give high resolution nmr spectra because of what kind of effects?

(*ii*) What are compounds containing no hydrogen atoms called?

(*iii*) What are *two* concentration dependant solvent-solute interactions?

(*iv*) The presence of what kind of impurities will lower the resolution of an nmr spectrum?

(*v*) What technique can often be used to remove labile protons in a sample?

Choose from:

(*a*) Paramagnetic

(*b*) Proton exchange

⟶

**SAQ 2.1c (cont.)**

(*c*) Chemical shift

(*d*) Aprotic

(*e*) Hydrogen-bonding

(*f*) Spin-active nucleus

(*g*) Line-broadening

(*h*) $D_2O$ exchange.

**Response**

The correct answers are:

(*i*) (*g*) Line broadening

(*ii*) (*d*) Aprotic

(*iii*) (*b*) and (*e*) Proton exchange and hydrogen-bonding

(*iv*) (*a*) Paramagnetic

(*v*) (*h*) $D_2O$ exchange

(*c*) and (*f*) were inappropriate to any of the questions.

**********************************

**SAQ 2.3a**

Which of the following best explains the term homogeneity as applied to magnetic fields?

(*i*) Homogeneity means how uniform the magnetic field is in the region of the sample.

(*ii*) Homogeneity means the magnetic field varies uniformly across the pole gap.

(*iii*) Homogeneity means the magnetic field is constantly varying as the nmr spectrum is being obtained.

(*iv*) A magnetic field is homogeneous when it is very stable.

**Response**

Statement (*i*) is the correct answer, homogeneity implies uniformity, constancy, lack of variation. Choice (*ii*) refers to a field gradient and while nmr imaging uses gradient magnetic fields high resolution nmr spectrometers certainly would not. Choice (*iii*) is incorrect as homogeneity does not mean constantly varying. Nevertheless, the strength of a magnetic field can vary yet its homogeneity remain constant. Similarly, choice (*iv*) must be incorrect as homogeneity is not the same phenomenon as stability.

************************************

**SAQ 2.3b**

Which of the statements below can be matched with:

(*i*) a permanent magnet

(*ii*) an electromagnet

(*iii*) a superconducting magnet?

*Note*: more than one statement may be appropriate for (*i*), (*ii*) or (*iii*)

(*a*) is used for high field nmr.

(*b*) is costly to run because of the price of electricity.

(*c*) is costly to run because of the price of liquid helium.

(*d*) provides stable magnetic field.

(*e*) is used for 60 MHz nmr spectrometers.

**Response**

For (*i*), a permanet magnet, statements (*d*) and (*e*) are appropriate.

For (*ii*), an electromagnet, statements (*b*), (*d*) and (*e*) are appropriate.

For (*iii*), a superconducting magnet, statements (*a*), (*c*) and (*d*) are appropriate.

Note that statement (*d*) applies to all the magnets. If they did not supply stable fields they would not be much use. Statement (*a*) could not apply to either permanent or electromagnets. In fact the term 'high field' nmr has been coined to refer to the high fields of superconducting magnets. Statement (*b*) could not, obviously, apply to a permanent magnet and a superconducting magnet, once set up, requires no further power for the magnet; it could if continually kept cold, run for ever (but keeping it cold does use energy). As neither permanent nor electromagnets require liquid helium statement (*c*) can apply only to superconducting magnets, and statement (*e*) is true for permanent and electromagnets although almost all modern 60 MHz instruments have permanent magnets.

**********************************

**SAQ 2.4a**

For each of the four statements decide which of the choice of answers is most suitable:

(*i*) This allows nmr spectra to be obtained at high temperature: (*a*) heating the sample in a steam bath; (*b*) warming the magnet; (*c*) using a variable temperature probe.

(*ii*) Variation of the current in these leads to a CW spectrum: (*a*) sweep coils; (*b*) receiver coils; (*c*) transmitter coils.

(*iii*) This is a common cause of not getting a sample to spin in the probe: (*a*) faulty nmr tube; (*b*) broken sweep coils; (*c*) too much sample.

**Response**

For (*i*) the correct response is (*c*), a variable temperature probe. If you heated the sample externally then it would be cooling as you recorded the spectrum, but, worse, the heat could disrupt the magnet temperature leading to loss of homogeneity and resolution. Similarly, warming the magnet is out of the question.

For (*ii*) (*a*) sweep coils is the correct answer, the receiver and transmitter coils are active all the time, but the sweep coils bring each part of the spectrum into resonance in turn.

Finally, a faulty tube (*a*) is the most likely cause of non-spinning samples in (*iii*). Nmr probes are pieces of very finely built equipment and any slight variation in the outside diameter of an nmr tube can stop it spinning. Other common causes of poor spinning are worn bearings and faulty air jets or spinners. The amount of sample in a tube should not affect the spinning of the tube, although in practice, most operators would never have more than 1 $cm^3$ of solution in a tube.

*********************************

**SAQ 2.4b**

A single resonance in an nmr spectrum was found to have spinning side bands. If the resonance was at $\delta$ = 2.0 ppm and the side bands at 2.3 and 1.7 ppm respectively and if the operating frequency of the instrument was 90 MHz, how fast was the tube spinning?

(*i*) 270 Hz

(*ii*) 90 Hz

(*iii*) 27 Hz

(*iv*) 54 Hz

**Response**

Answer (*iii*) is correct; here's how to work it out. The side bands are symmetrically placed 0.3 ppm either side of the main spectral line. Thus as

$$1 \text{ ppm} \equiv 90 \text{ Hz at } 90 \text{ MHz}$$

$$0.3 \text{ ppm} \equiv 0.3 \times 90$$

$$= 27 \text{ Hz.}$$

If you thought (*i*) to be correct you probably had some arithmetical error. (*ii*) is not related in any way to the observed displacement of the side bands.

Finally, (*iv*) might be a plausible answer as it is the frequency difference between the side bands, but the actual spinning speed is frequency difference between the real spectral line and either one of its side bands.

**********************************

**SAQ 2.5**

Which of the following statements about radiofrequency sources and detectors are correct?

Indicate your answer by circling T (true) or F (false).

(*i*) A typical Rf power level in CW nmr is 100 watts.

T / F

(*ii*) In nmr the term 'saturation' means that no more sample will dissolve in an nmr solvent.

T / F

(*iii*) Nmr detectors are not very sensitive.

T / F

⟶

**Response**

Statement (*i*) is false; such a high level of power is appropriate for FT pulsed nmr, but for CW nmr about 1 watt is the usual power level.

Statement (*ii*) is incorrect; saturation in nmr spectroscopy means that the nmr signals are being distorted by too high an Rf power level.

Alas statement (*iii*) is also false. Nmr detector systems have to sense very small (1 millivolt and less) changes in signal levels. So they have to be quite sensitive. Remember, though, that the signal has to be amplified and this leads to electronic noise.

***********************************

**SAQ 2.7a**

Each of the following statements contains a word or phrase that is not necessarily correct. Pick that word or phrase out and substitute a correct word or phrase.

(*i*) After being detected, the nmr signal is automatically sent to a computer for processing.

(*ii*) Chart recorder output of an nmr spectrum occurs more quickly than an oscilloscope output.

(*iii*) Overall, an nmr spectrometer is quite a simple instrument.

**Response**

The word 'automatically' is not necessarily true in statement (*i*).

While computer processing of nmr data can be necessary, most routine instruments do not have computers associated with them. A substitute word might be 'sometimes'.

Concerning statement (*ii*), the word 'quickly' should be replaced by 'slowly'. The quality of CW nmr spectra is markedly dependant on how fast the spectrum is run; the slower the better usually. With an oscilloscope we are not usually too worried about the resolution of the spectrum, more with whether there is a spectrum at all and with adjustments to the major controls.

The word 'simple' in statement (*iii*) would not be considered correct by most students or nmr experts. Nmr instruments are really quite complex although a number of manufacturers have tried to produce 'user friendly' spectrometers that can be used with a minimum of training. These instruments are normally limited in their capabilities.

***********************************

**SAQ 2.7b** Make your own sketch of the basic components of a CW nmr spectrometer and its probe.

**Response**

Your sketch should look like Fig. 2.7a and Fig. 2.4a. If not, try again or claim you are related to Pablo Picasso!

***********************************

**SAQ 3.2**

The carbon-13 spectrum for diethyl ether gives two peaks, at $\delta_C$ = 14.1 ppm and $\delta_C$ = 64.8 ppm. In the off-resonance spectrum the first peak splits into a quartet, while the second becomes a triplet. Match these data with the structure.

**Response**

Diethyl ether has only two structurally different carbon atoms, the $CH_3$ and the $CH_2$ carbons. Consulting Fig. 3.1d, the carbon-13 correlation chart tells us that the $CH_2$ carbon, being bonded to oxygen should resonate somewhere between 40 and 70 ppm while the methyl carbon would come in the 10–30 ppm region. Thus the peak at 64.8 ppm belongs to the $CH_2$ and at 14.1 ppm to the $CH_3$.

The splitting pattern in the off-resonance spectrum confirms these assignments because the peak at 14.1 ppm being split into a quartet must belong to a $CH_3$ (remember the $2nI + 1$ rule), and similarly the peak at 64.8 ppm must come from a $CH_2$ as it is split into a triplet.

If you matched these resonances/carbons the other way round then I suggest revision of Part 1 and the start of this Part. You must have a thorough grasp of simple systems like diethyl ether before you can tackle more complex ones.

**********************************

**SAQ 3.3**

Fig. 3.3d and 3.3e show the proton and carbon-13 nmr spectra for a pure compound. Determine the structure of the compound and assign all the resonances as far as possible.

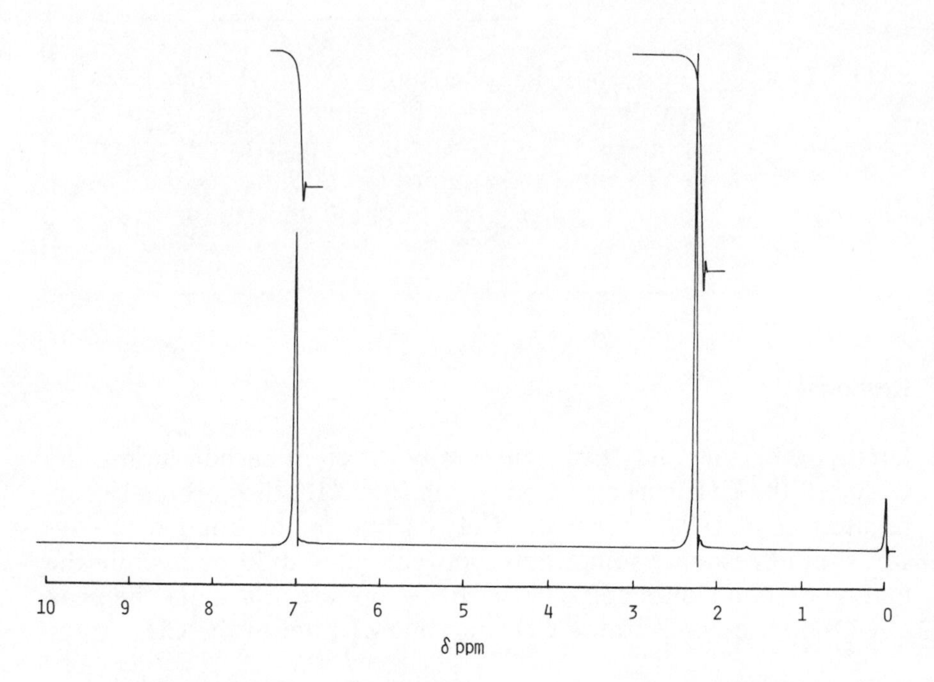


**Fig. 3.3d.** *Proton nmr spectrum for SAQ 3.3 ($CDCl_3$ solution)*

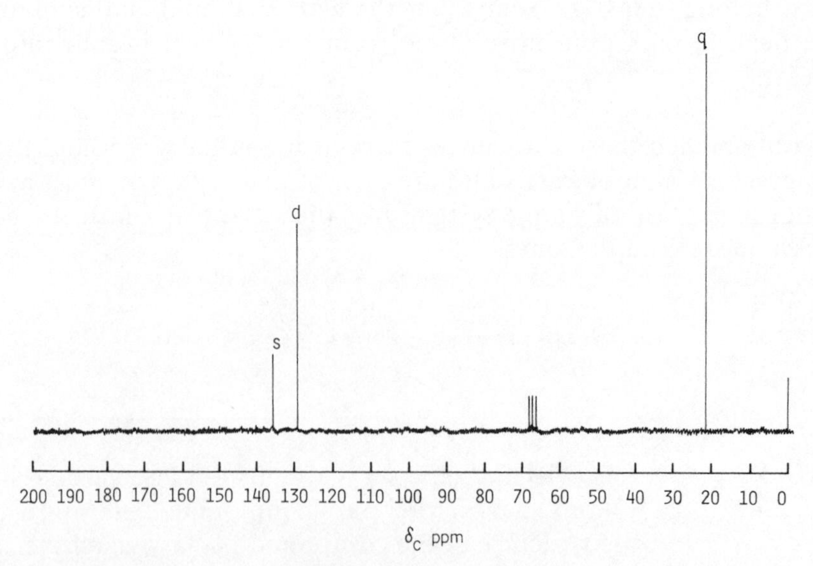


**Fig. 3.3e.** *Carbon-13 nmr spectrum for SAQ 3.3 ($CDCl_3$ solution)*

**Response**

Starting with the proton spectrum, Fig. 3.3d, you should have seen two resonances at $\delta$ = 2.3 ppm and 7.1 ppm with no coupling apparent. Using correlation table (Fig. 3.1a and/or the correlation chart, Fig. 3.1c, you should have reasoned that there must be alkyl protons (2.3 ppm) attached to an aromatic ring (7.1 ppm). Checking the integration curves you would have found a ratio of 3 : 2 for the alkyl protons to aromatic protons. This is rather a difficult ratio to fit if there is only one alkyl group substituent, so perhaps a ratio of 6 : 4 is more realistic.

Turning to the carbon-13 spectrum, Fig. 3.3e, there are three resonances at $\delta_C$ = 20.9, 129.0 and 134.5 ppm. Using the appropriate carbon-13 correlation chart you should have confirmed that this compound is aromatic (the resonances at 129.0 and 134.5 ppm) with alkyl group substitution (20.9 ppm). The off-resonance data tells us that the alkyl substituent must be a methyl group as the resonance at $\delta_C$ = 20.9 ppm splits into a quartet. The resonance at $\delta_C$ = 134.5 ppm remains as a singlet—so it belongs to the carbon to which the methyl group is bonded—while the resonance at 129.0 ppm splits to a doublet telling you that this must be CH in the aromatic ring.

So combining our findings we must have the following structural features:

$\delta_C$=20.4 ppm ------- $CH_3$ ------- $\delta$ =2.3 ppm
$\delta_C$=134.5 ppm ------- C
$\delta_C$=129.1 ppm ------- HC  CH ------- $\delta$ =7.1 ppm

This gives a proton ratio of 3 : 2, and two different proton and carbon-13 resonances. So the full structure of this compound must be:

$CH_3$

$CH_3$

1,4-dimethylbenzene
(p-xylene)

Only this symmetrical structure would give such simple spectra. Any other isomers of 1,4-dimethylbenzene would have a different number of types of carbon and hydrogen atoms and so give more complex spectra. We shall return to the problem of symmetry in section 3.7.

***********************************

**SAQ 3.4**

Choose for each of the molecular formulae below the correct number of double bond equivalents.

(*i*) $C_4H_8$ 1 or 2 or 4

(*ii*) $C_6H_6$ 2 or 4 or 6

(*iii*) $C_5H_4O_2$ 3 or 4 or 5

(*iv*) $C_5H_5N$ 4 or 5 or 6

**Response**

You should have applied the rules for calculating double bond equivalents discussed in section 3.4.

For $C_4H_8$ the saturated equivalent would be $C_4H_{10}$,

∴ 2H difference

∴ 1 double bond equivalent for (*i*).

For $C_6H_6$ the saturated equivalent is $C_6H_{14}$,

∴ 8H difference

∴ 4 double bond equivalents for (*ii*); 1 for the aryl ring and 3 for the double bonds of the ring.

For $C_5H_4O_2$ this must be converted to the CH equivalent by replacing $CH_2$ for every O atom. So the formula becomes $C_7H_8$. The saturated equivalent is $C_7H_{16}$,

∴ 8H difference,

∴ 4 double bond equivalents for (*iii*)

Similarly for $C_5H_5N$ the N must be replaced by CH so it becomes $C_6H_6$ and applying the reasoning for (*ii*) there must be 4 double bond equivalents for formula (*iv*).

***********************************

**SAQ 3.5**

A compound of molecular formula $C_8H_8O$ has the following proton and carbon-13 spectral data. Deduce the structure of the compound.

Proton spectrum ($\delta$ ppm): 2.5 (3H); 7.4–8.0 (5H)

Carbon-13 spectrum ($\delta_C$ ppm): 26.3 (q); 128.2 (d); 128.4 (d); 132.9 (d); 137.1 (s); 197.6 (s)

The number of protons for each resonance is shown in brackets in the proton data as is the off-resonance splitting for the carbon-13 data, (s – singlet, d – doublet, q – quartet).

**Response**

Starting with the molecular formula, $C_8H_8O$, and using the rules in section 3.4 you should have calculated the number of double bond equivalents to be 5. ($C_8H_8O \rightarrow C_9H_{10} \rightarrow C_9H_{20}$, 10 difference, so 5 double bond equivalents). This large number of double bond equivalents in such a low carbon formula strongly suggests an aromatic ring as part of the structure.

Using the proton chemical shift correlation charts/tables and the given number of protons you should have deduced the presence of a mono substituted benzene ring ($\delta$ = 7.4–8.0 ppm, 5H) and a methyl group $\delta$ = 2.5 ppm, 3H).

Using the carbon-13 data the above deductions are confirmed, the resonance at 26.3 ppm splitting to a quartet being characteristic of a methyl group and the set of resonances at 128–137 ppm being from four different types of aromatic carbon atom. In addition, the carbon-13 data tells us that there is a carbonyl carbon resonating at 197.6 ppm. A carbon oxygen double bond is then the last of the five double bond equivalents, the aromatic ring having the other four.

Summarising all this information we have:

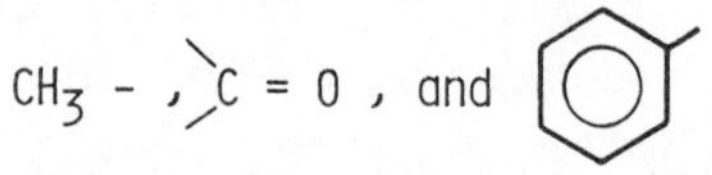

$CH_3-$ , $>C=O$ , and

These fragments can fit together in only one way, so the structure of the compound must be:

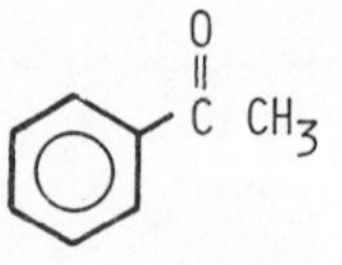

$C_6H_5-C(=O)CH_3$

acetophenone

************************************

**SAQ 3.7**

The proton and carbon-13 spectra of a compound of unknown structure containing only C, H and O are shown in Fig. 3.7c and 3.7d respectively. Deduce the structure of the compound.

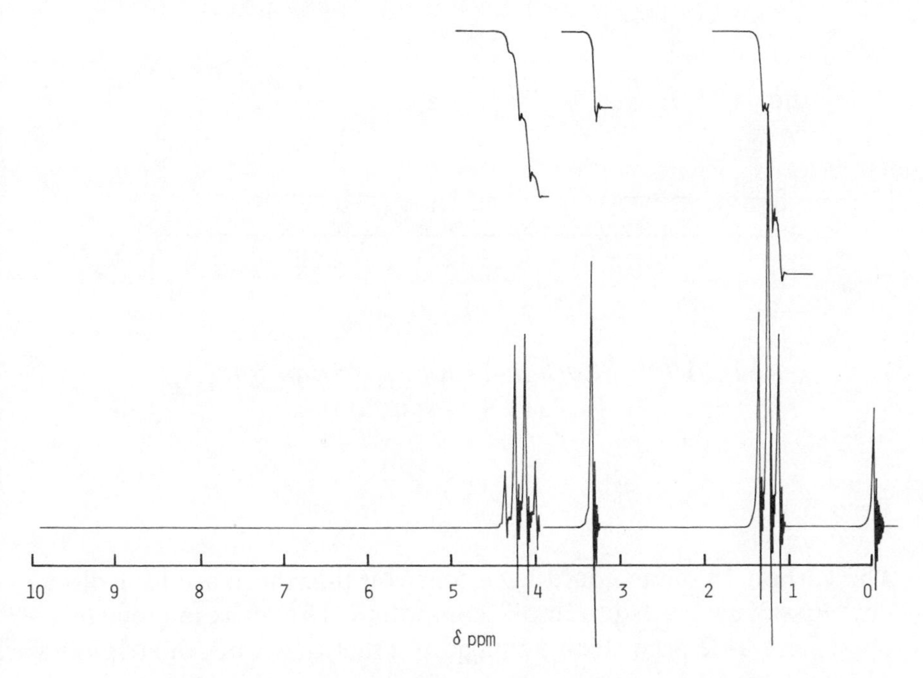


**Fig. 3.7c.** *Proton nmr spectrum for SAQ 3.7 ($CDCl_3$ solution)*

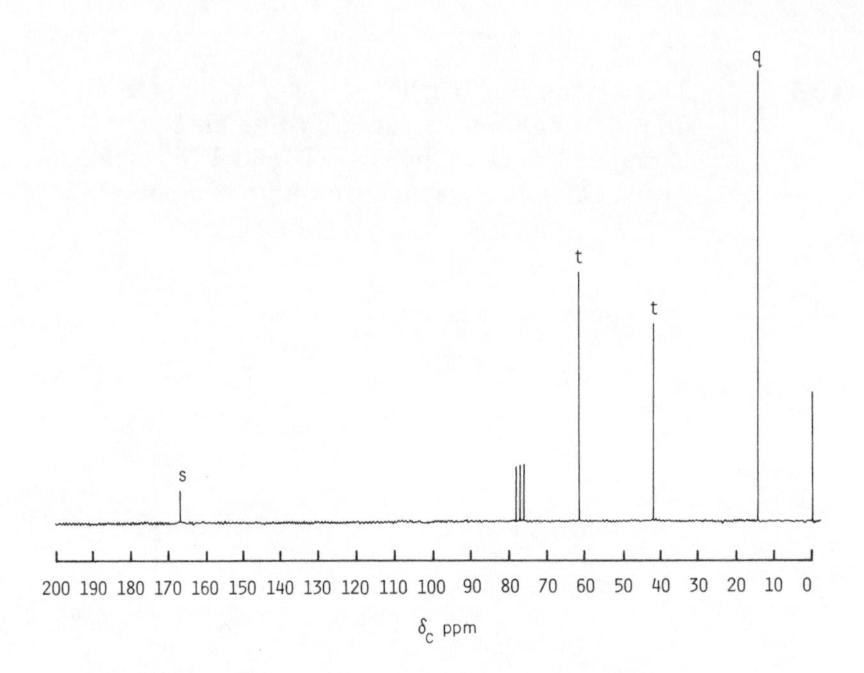


**Fig. 3.7d.** *Carbon-13 nmr spectrum for SAQ 3.7 ($CDCl_3$ solution)*

**Response**

The carbon-13 data should have told you that there are four different types of carbon atom in this compound. The off-resonance quartet at $\delta_C = 14.2$ ppm strongly suggests a methyl group, the triplets at 41.6 ppm and 61.3 ppm suggest methylene groups, the latter probably being bonded to an oxygen atom, and the singlet at 166.7 ppm suggesting a carbonyl of some type. (Remember you are told there is only carbon, hydrogen and oxygen in this material).

Thus in terms of molecular fragments the carbon-13 spectrum should have given you:

$$CH_3-, \quad -CH_2-, \quad -CH_2-O-, \quad C=O$$

The proton spectrum substantiates these features and helps us to put the structure together. Perhaps the most obvious part of this spectrum is the triplet (1.2 ppm) and quartet (4.2 ppm) which are clearly coupled to one another. These peaks suggest an ethyl group bonded to oxygen as part of an ester group. Remember to check your assignments with the correlation tables.

There is also a singlet resonance at $\delta$ = 3.4 ppm which in terms of chemical shift might suggest a group bonded to oxygen. Also, back in the carbon-13 spectrum there was a triplet at $\delta_C$ = 41.6 ppm which was not fully assigned. As all the other proton and carbon-13 resonances have been matched then this resonance and the proton one at 3.4 ppm probably come from the same methylene group.

The integration of the proton spectrum should have given you a ratio of 3:1:2 for the peaks at 1.2, 3.4 and 4.2 ppm respectively. Now the 3:2 ratio fits for an ethyl group, but the resonance at 3.4 ppm you know to be from two protons (on a methylene group). So you should have deduced that there are *two* ethyl groups, the true ratio of peak areas being 6:2:4.

If there are two ethyl groups there are two ester groups, so you should have deduced the following molecular fragments:

$$CH_3CH_2OCO-, \quad CH_3CH_2OCO-, \quad \text{and} \quad -CH_2-$$

The only way these can be put together is as diethylmalonate:

$$CH_3CH_2OCO-CH_2-COOCH_2CH_3$$

So the resonances at $\delta_C$ = 41.6 ppm and $\delta$ = 3.4 ppm were due to the $CH_2$ bonded to two carbonyl groups, not to an oxygen atom.

This was quite a hard problem and its solution depended on correctly using the integration information. As we have seen with diethyl ether, 1,4-dimethylbenzene and 2,4-dimethylpentan-3-one symmetry is an important simplifying phenomenon in nmr spectra, yet one which can make interpretation quite difficult.

**********************************

**SAQ 3.8**

Fig. 3.8d and 3.8e show the proton and carbon-13 nmr spectra of a compound of unknown structure which contains only C, H and O. Deduce the structure of this compound and explain all the resonances.

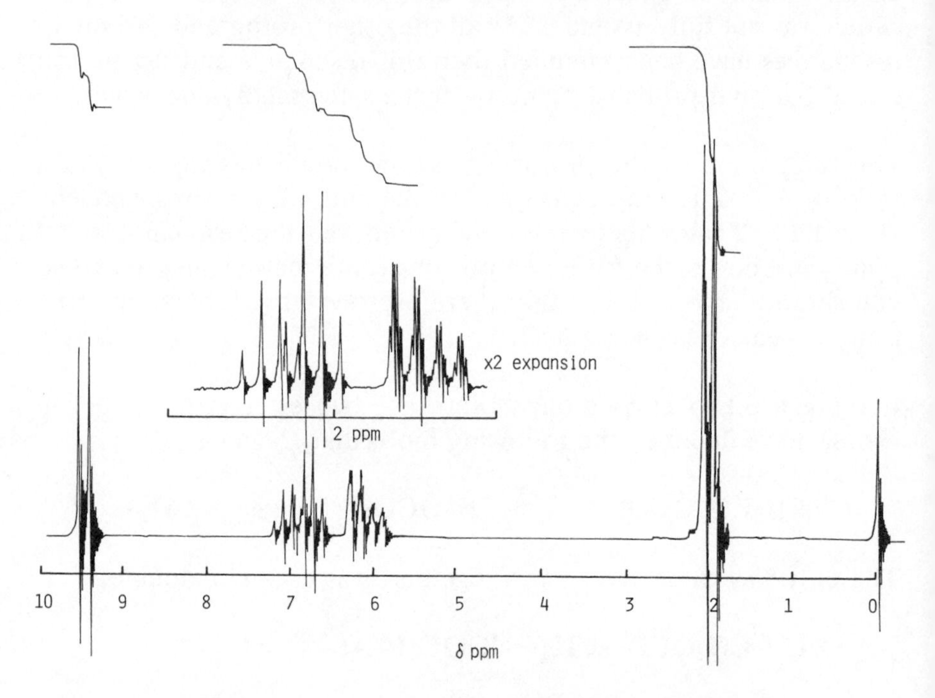


**Fig. 3.8d.** *Proton nmr spectrum for SAQ 3.8 ($CDCl_3$ solution)*

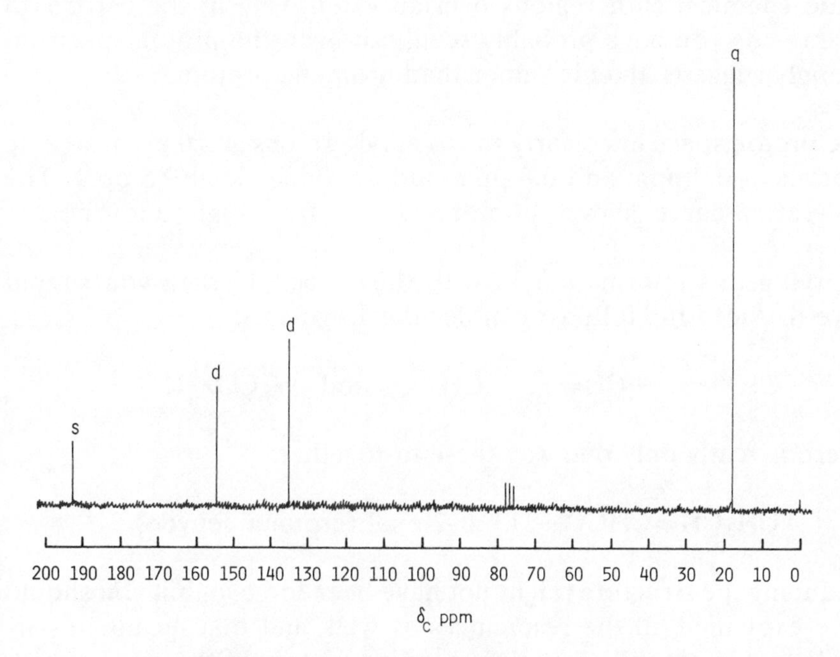


**Fig. 3.8e.** *Carbon-13 spectrum for SAQ 3.8 ($CDCl_3$ solution)*

**Response**

This problem is probably the most complex in this part of the Unit. So if you have made all the correct deductions and assignments you should be pleased with yourself. If not, don't worry, there is plenty of scope for more practice on the problems of section 3.10.

Starting again with the carbon-13 spectrum (remember the last heuristic: choose the simpler data first) there are four resonances, so four different types of carbon atom. Combining chemical shifts and off-resonance data you should have concluded that there is a methyl group attached to carbon (18.2 ppm, quartet), aldehyde group (193.4 ppm, doublet), and two other CH's probably alkenic (134.9 ppm, doublet and 153.7 ppm, doublet). Why are these latter two resonances not part of an aromatic ring? After all the alkenic and aromatic chemical shift regions overlap extensively in the carbon-13 spectra. As you have probably seen, however, the proton spectrum strongly suggests alkenic rather than aromatic protons.

The proton spectrum clearly shows alkyl protons at 2.0 ppm, alkenic protons at 6.0 ppm and 6.8 ppm and an aldehyde at 9.5 ppm. The integration curve gives a ratio of 3 : 1 : 1 : 1 from high to low field.

So, tying this information in with the carbon-13 data you should have deduced the following molecular fragments:

$CH_3-$, $-CH=$, $-CH=$, and $-CO-H$

There is really only one way these fit together:

$CH_3CH{=}CHCO-H$ but-2-enal (crotonaldehyde)

Deducing the structure might not have been too bad, but you should have explained all the resonances as well, and that means in particular explaining the complex coupling pattern in the proton spectrum. Remember one of the heuristics of interpretation was that all the data should be explained.

In this instance you could work back from the structure to work out the splitting pattern for the alkenic protons. I will label them $H_A$ and $H_X$ for convenience:

$CH_3CH_A{=}CH_XCO-H$

Now $H_A$ is coupled strongly to the methyl group so it will be a quartet from that interaction. A coupling constant of about 7 Hz might be expected. $H_A$ is also coupled to $H_X$, a doublet with a coupling of about 12–15 Hz resulting. Overall a doublet of quartets is expected. Fig. 3.8f shows this and allows us to assign the peaks at $\delta$ = 6.8 ppm to $H_A$.

Similarly, the pattern for $H_X$ can be worked out. A doublet results from coupling to $H_A$, each of these lines becomes a doublet from coupling to aldehyde proton, and as seen in the scale expanded part of Fig. 3.8d, each of those lines becomes a quartet from long range coupling to the methyl group (Fig. 3.8g shows this). The variation in intensities is not quite what you would have expected, but to explain why is beyond the scope of this work.

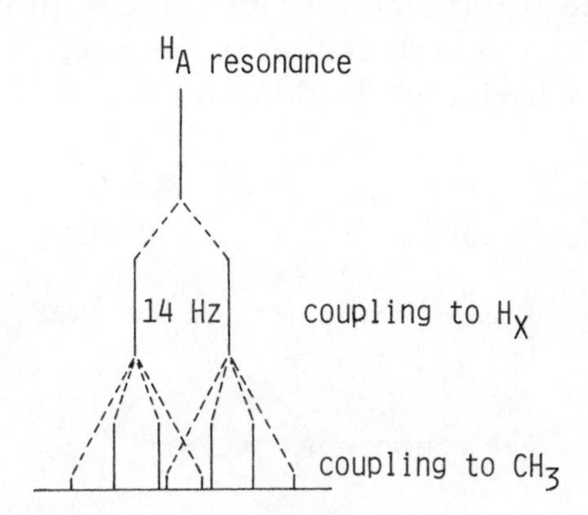


**Fig. 3.8f.** *Coupling pattern for $H_A$ in but-2-enal for SAQ 3.8*

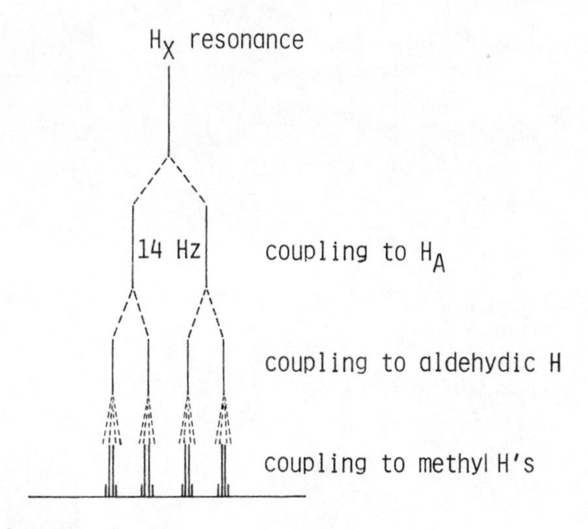


**Fig. 3.8g.** *Coupling pattern for $H_X$ in but-2-enal for SAQ 3.8*

**SAQ 3.9a to SAQ 3.9f: 6 problems**

Deduce the structure for each compound from the proton and carbon-13 spectra and other data that are provided. You should also explain your interpretation of the spectra as far as possible. The problems are loosely graded with the fairly easy ones first.

**SAQ 3.9a**

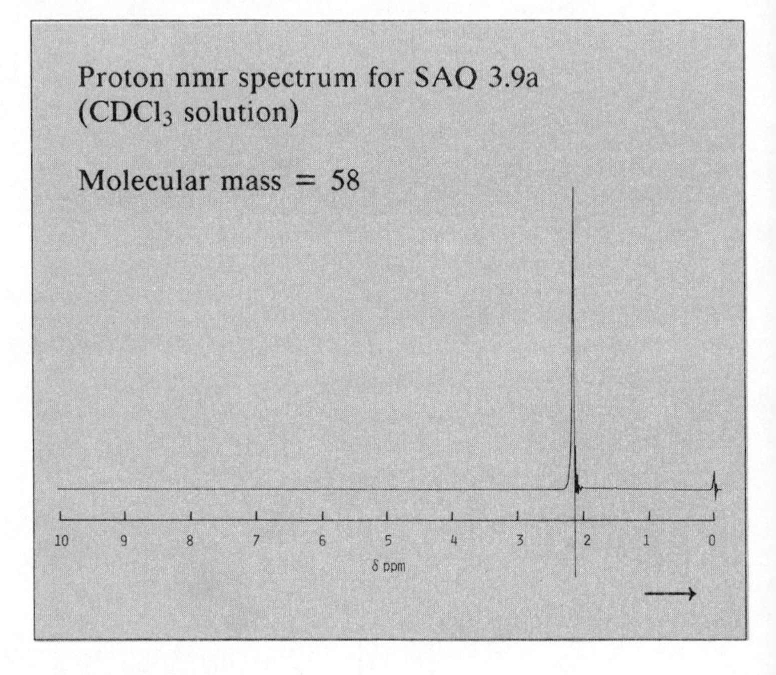

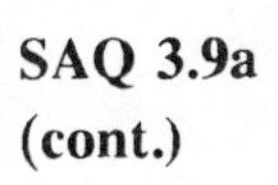

**SAQ 3.9a (cont.)**

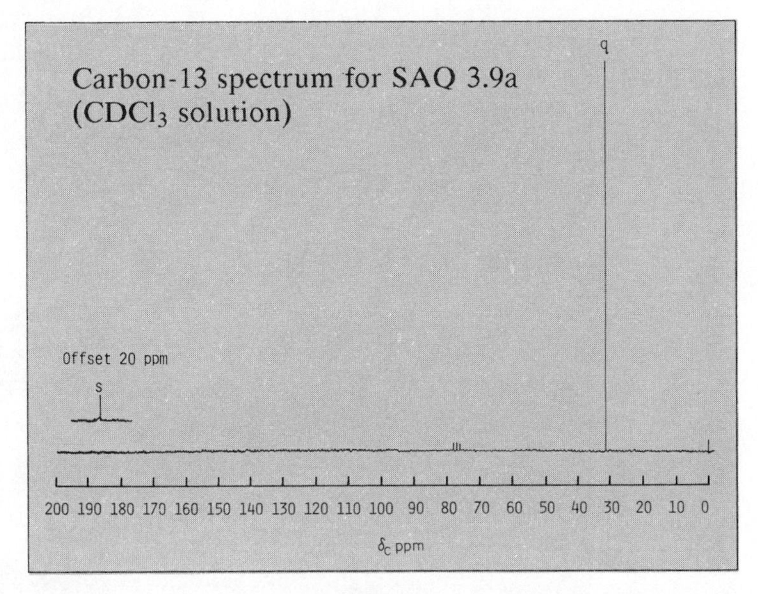


**Response**

This, hopefully, was an easy start to the problem set. Taking the proton spectrum first, there appears to be only one type of hydrogen in this compound, its resonance being at $\delta = 2.1$ ppm. From the carbon-13 spectrum the structure has only two different types of carbon atom, ignoring the triplet resonance at $\delta_C = 77$ ppm which comes from the solvent, $CDCl_3$. Consultation of the carbon-13 correlation chart (Fig. 3.1d) should have told you that the weak resonance at 208 ppm comes from a ketonic carbon atom, while the off-resonance quartet at 31 ppm must belong to a methyl carbon. The proton resonance at 2.1 ppm is typical of protons on a methyl group bonded to a carbonyl (check with the correlation tables and chart) thus confirming the carbon-13 deductions. As there are no other carbon or hydrogens evident the most likely structure must be:

$CH_3COCH_3$ propanone (acetone)

The molecular mass, 58, is the same as that given on the proton spectrum.

***********************************

**SAQ 3.9b**

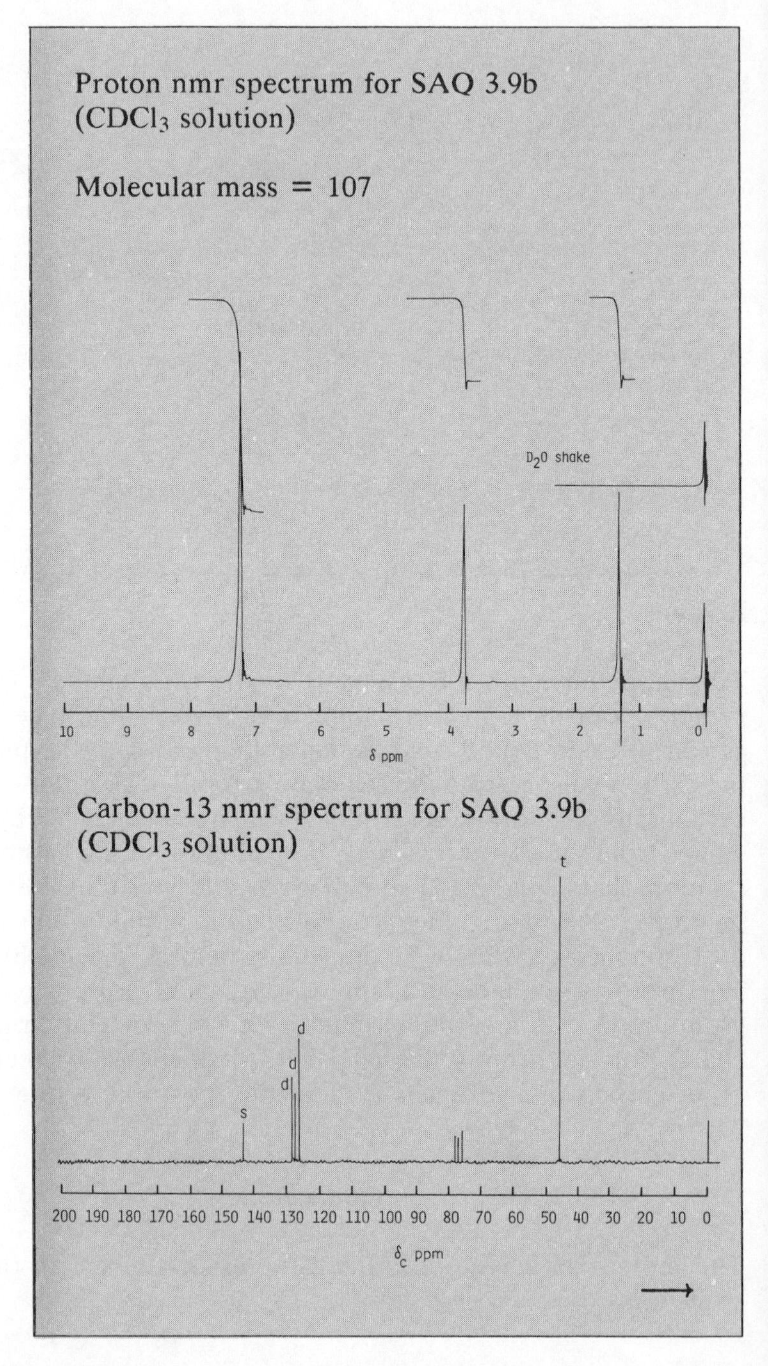
Proton nmr spectrum for SAQ 3.9b
(CDCl3 solution)
Molecular mass = 107
D2O shake
10 9 8 7 6 5 4 3 2 1 0
δ ppm
Carbon-13 nmr spectrum for SAQ 3.9b
(CDCl3 solution)
t
d
d
d
s
200 190 180 170 160 150 140 130 120 110 100 90 80 70 60 50 40 30 20 10 0
δc ppm
⟶

**Response**

You should have spotted the clue in the molecular mass of 107. An odd molecular mass suggests an odd number of nitrogen atoms are present in the structure. Looking at the proton spectrum you should have noticed that one of the three resonances, at 1.2 ppm, disappears upon shaking the sample with $D_2O$. The integration ratio of 2:2:5 (from high to low field) suggests two protons for the resonance at 1.2 ppm and so this could well be from an $-NH_2$ group.

The peak at 7.2 ppm suggests a monosubstituted benzene ring and the presence of four peaks in the aromatic region of the carbon-13 spectrum, bears out this idea. There is also a triplet at $\delta_C = 46$ ppm suggesting a $CH_2$ bonded perhaps to nitrogen. (Check with your correlation tables and charts).

Thus the spectra have suggested the following: monosubstituted benzene, $CH_2$ and $NH_2$. The only possible structure is then:

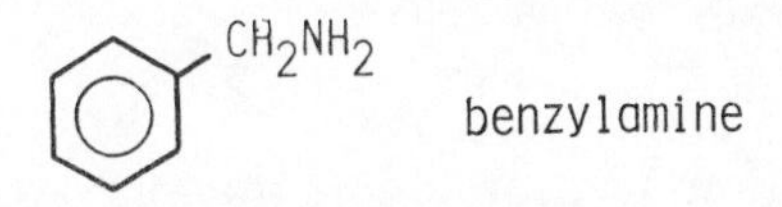


Perhaps that was a bit tricky with the nitrogen atom thrown in, but remember that you are just starting with this type of problem and your skills will develop.

**********************************

**SAQ 3.9c**

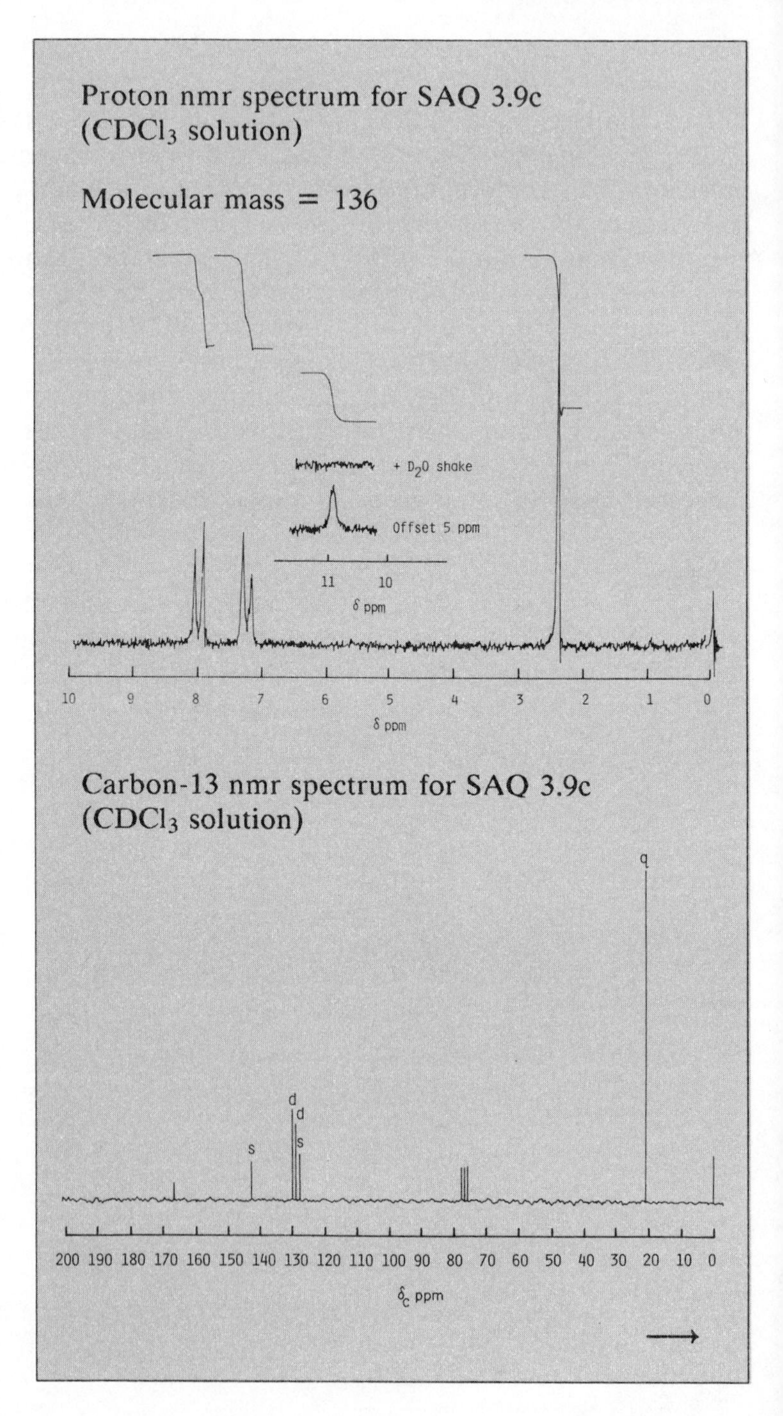
Proton nmr spectrum for SAQ 3.9c
($CDCl_3$ solution)
Molecular mass = 136
+ D2O shake
Offset 5 ppm
11
10
δ ppm
10 9 8 7 6 5 4 3 2 1 0
δ ppm
Carbon-13 nmr spectrum for SAQ 3.9c
($CDCl_3$ solution)
q
d
d
s
s
200 190 180 170 160 150 140 130 120 110 100 90 80 70 60 50 40 30 20 10 0
δc ppm

**Response**

You should have spotted that this structure has an aromatic group in it from the doublet resonances at 7 and 8 ppm in the proton spectrum and the set of four peaks between 128 and 144 ppm in the carbon-13 spectrum.

An off-resonance quartet at 20 ppm suggests a methyl group and this is confirmed by the singlet resonance at 2.4 ppm in the proton spectrum. The precise chemical shift in both cases suggests that the methyl could be bonded to the aromatic ring. The proton spectrum also shows a $D_2O$ exchangeable resonance at $\delta = 11$ ppm which is highly characteristic of a carboxylic acid proton. The carbon-13 spectrum backs this up with a singlet resonance at 168 ppm.

So the data has been interpreted to suggest $CH_3$, $CO_2H$, and an aromatic ring. Assuming the aromatic group to be a disubstituted benzene we have a molecular mass of 136, as is given in the problem. We now have to decide which of the three isomers shown below our compound is:

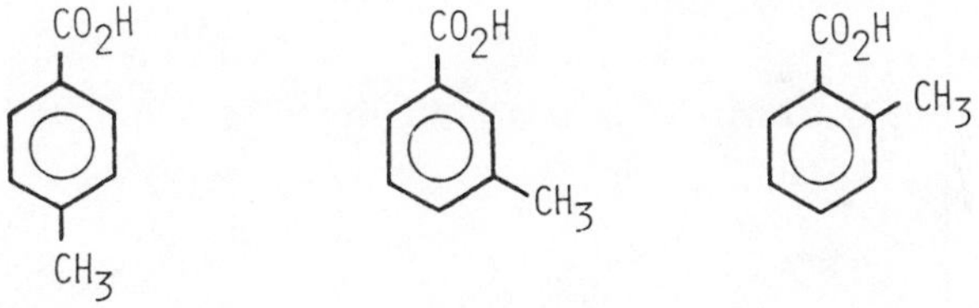

$CO_2H$ $CO_2H$ $CO_2H$

$CH_3$

$CH_3$

$CH_3$

Possibly the easiest way to decide in this case is to count the number of different types of carbon atom in the aromatic ring of each of the isomers. From left to right it is 4, 6 and 6. As we see only four lines in the carbon-13 spectrum in the aromatic region we could infer that the first isomer, 4-methylbenzoic acid, is the correct structure. The proton spectrum backs this assertion because there are only two different types of hydrogen in the aromatic ring which coupling to one another give the apparent doublet of doublets. The other structures would give much more complex proton nmr spectra in the aromatic region.

***********************************

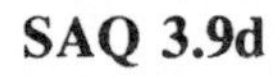

**SAQ 3.9d**

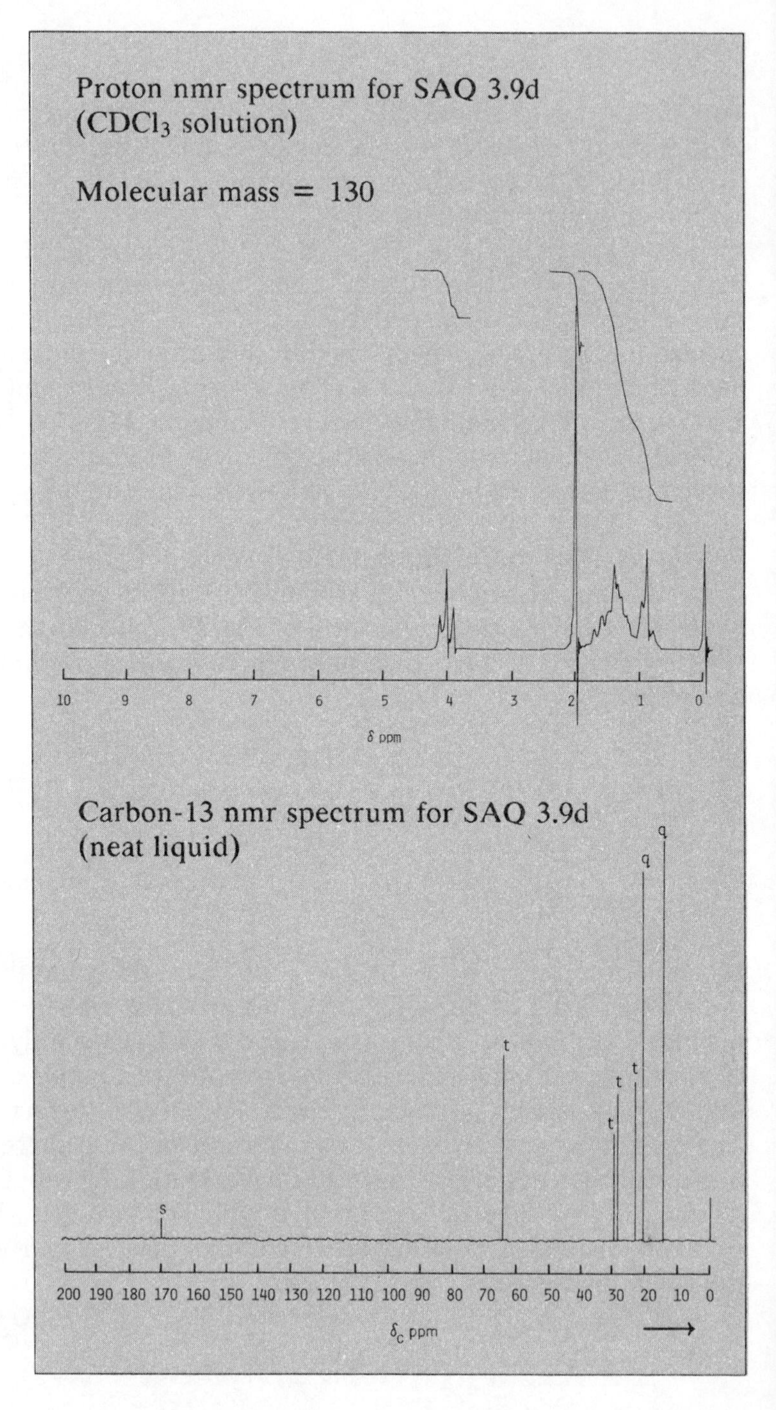
Proton nmr spectrum for SAQ 3.9d
(CDCl3 solution)
Molecular mass = 130
10 9 8 7 6 5 4 3 2 1 0
δ ppm
Carbon-13 nmr spectrum for SAQ 3.9d
(neat liquid)
s
t
t
t
t
q
q
200 190 180 170 160 150 140 130 120 110 100 90 80 70 60 50 40 30 20 10 0
δC ppm

**Response**

You might have appreciated the power of carbon-13 and proton nmr spectroscopy working together in this problem. The proton specrum, while offering a number of clues, shows a confusing set of overlapping resonances in the 1.2 ppm region. The difficulty in interpretation is lifted when the carbon-13 spectrum is consulted because a clear set of alkyl carbon resonances is seen between 10 and 30 ppm.

In fact the carbon-13 spectrum shows that there are seven different types of carbon atom including a carbonyl carbon resonating at 170 ppm. The precise chemical shift suggests an ester group. The triplet at 65 ppm suggests $CH_2$ bonded to an electronegative element, probably oxygen. The two quartets at 15 and 20 ppm must come from methyl group carbon atoms while the other three triplets suggest three methylene groups. So, in total, we have 4 $CH_2$'s, 2 $CH_3$'s and C=O with a combined mass of 114. As the actual molecular mass is given as 130 there must be another oxygen atom.

The proton spectrum confirms the above deductions and helps us to assemble the structure. The triplet at $\delta$ = 4.0 ppm must be a $CH_2$ adjacent to oxygen; the sharp singlet at 2.0 ppm suggests a $CH_3$ next to a carbonyl and the overlapping resonances between 1 and 2 ppm must belong to an alkyl chain.

Thus we have:

$CH_3COOCH_2CH_2CH_2CH_2CH_3$ n-pentylethanoate (amyl acetate)

Well done if you got the last two problems correct first time, but don't give up if you didn't. You can always have a break and try them afresh.

************************************

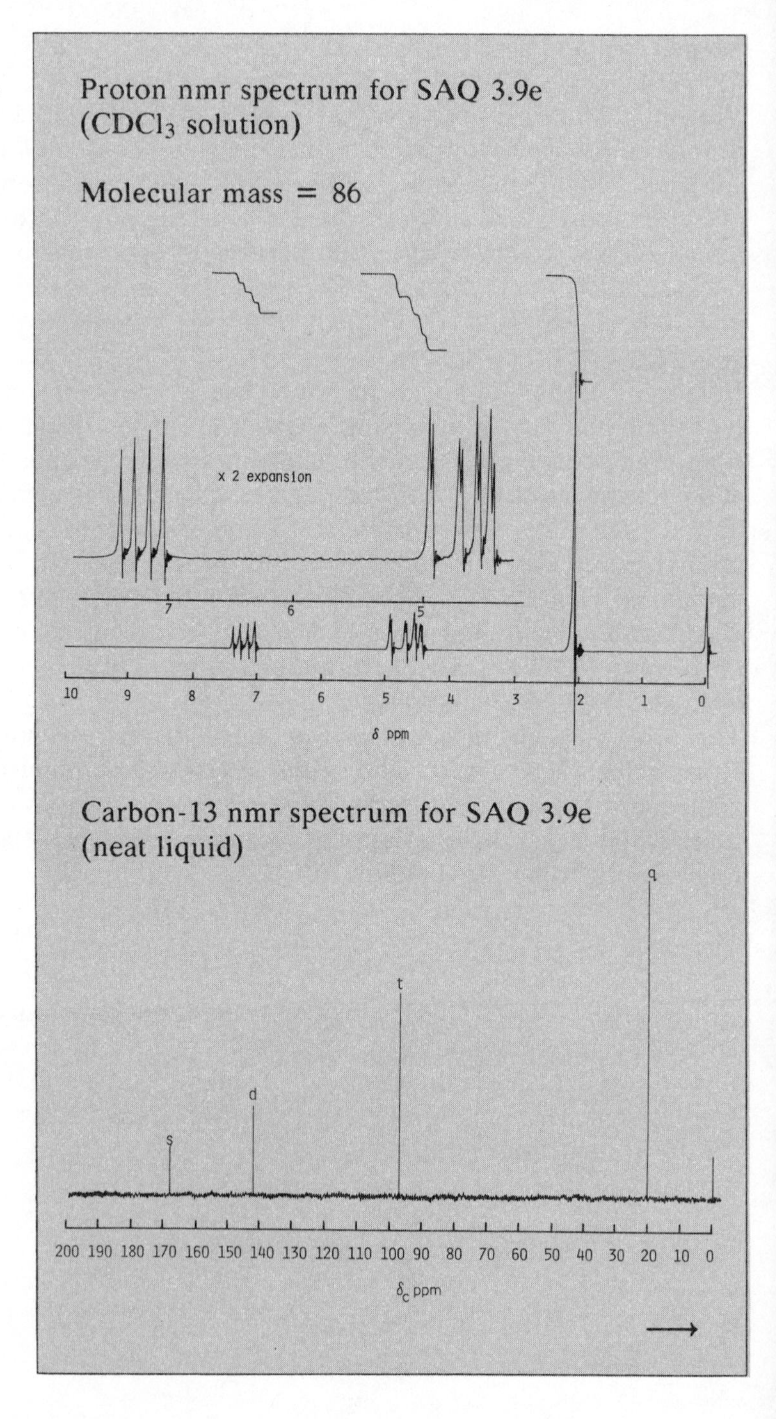

Proton nmr spectrum for SAQ 3.9e
(CDCl3 solution)
Molecular mass = 86
x 2 expansion
7
6
5
10
9
8
7
6
5
4
3
2
1
0
δ ppm
Carbon-13 nmr spectrum for SAQ 3.9e
(neat liquid)
q
t
d
s
200 190 180 170 160 150 140 130 120 110 100 90 80 70 60 50 40 30 20 10 0
δc ppm

**Response**

You should have noted that there are four different types of carbon atom in this compound from the four resonances of the carbon-13 spectrum. The off-resonance data should have told you that there is a carbon atom with no protons attached (168 ppm) the chemical shift suggesting a carbonyl carbon, possibly an ester carbon. A doublet at 142 ppm suggests an alkenic CH (the material cannot be aromatic as the molecular mass is too low to support an aromatic group plus the other functionality that is evident). The triplet at 97 ppm suggests a $CH_2$ group and from the correlation chart (Fig. 3.1d) a possible assignment would be to a $CH_2$ in a vinyl ether or ester, $-O-C{=}CH_2$. The quartet at 20 ppm strongly suggests a methyl carbon, possibly attached to a carbonyl. So from the carbon-13 data we have the following: $CH_3-$, $CH_2{=}$, $CH{=}$, $>C{=}O$.

The proton spectrum shows alkenic protons at 4.7 and 7.2 ppm strongly coupled to one another. The ratio of 2 : 1 supports the idea of $CH_2{=}CH-$. The singlet resonance at 2.05 ppm should have suggested a $CH_3$ group bonded to unsaturation, possibly a carbonyl.

Putting all the evidence so far together this structure must have $CH_3-$, $CH_2{=}CH-$, and $C{=}O$, totalling 70 mass units. So the missing amount must be an oxygen atom (why is it not $-NH_2$ ?). The only way of putting this molecular jigsaw together is:

$CH_3CO-O-CH{=}CH_2$ vinyl ethanoate (vinyl acetate)

The coupling pattern for the alkenic proton resonances is as follows:

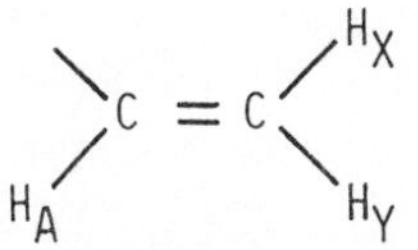

$H_X$

C = C

$H_A$ $H_Y$

The proton whose resonance shows four lines centred at 7.2 ppm, let us call it $H_A$, is split by the proton which is trans to it on the double bond (call it $H_X$) into a doublet. In turn the proton cis to $H_A$ (call it $H_Y$) splits both lines into a doublet. Hence the four observed lines is a doublet of doublets. The four doublets centred at 4.7 ppm arise from $H_X$ and $H_Y$ which have separate resonances being split into doublets by $H_A$ and into further doublets by coupling to one another. That is $H_X$ splits $H_Y$ and vice versa. The actual coupling constants turn out to be $J_{AX}$ = 13 Hz, $J_{AY}$ = 6.6 Hz, and $J_{XY}$ = 2 Hz. (Check with the correlation table).

You are right, these are getting tougher. Well done indeed if you got all of the points noted above.

**********************************

**SAQ 3.9f**

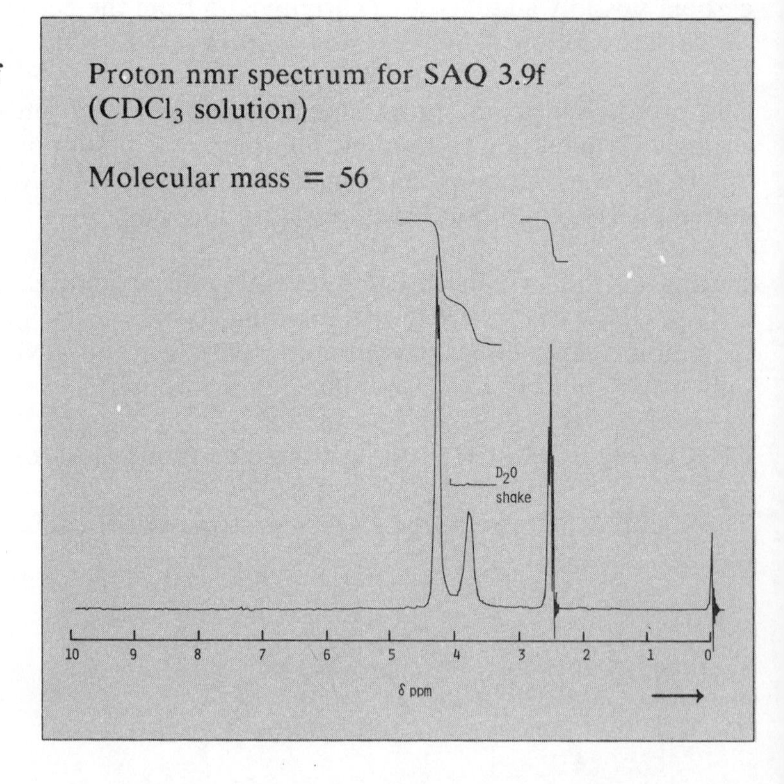


Proton nmr spectrum for SAQ 3.9f ($CDCl_3$ solution)

Molecular mass = 56

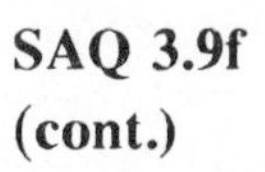


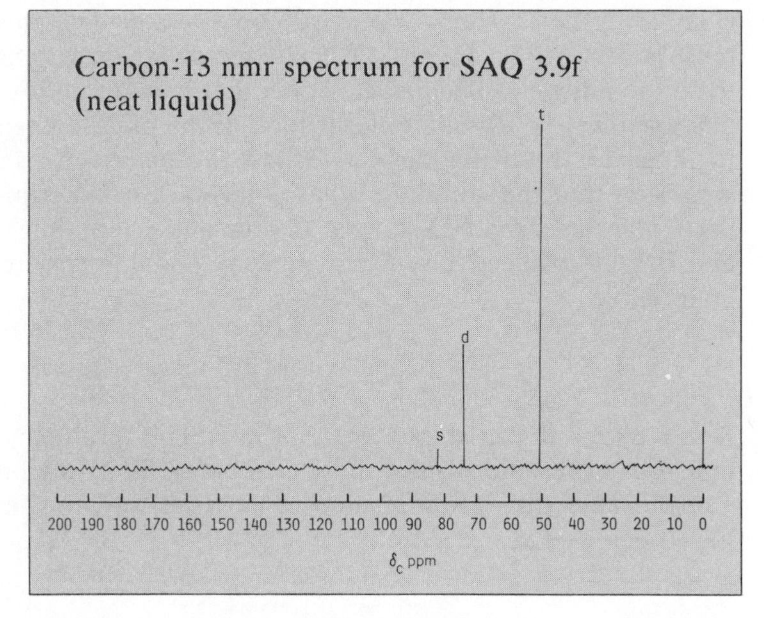


**Response**

You might have found this problem a bit tricky, despite the low molecular mass, 56, which suggests few atoms in the structure. The carbon-13 spectrum and its off-resonance information should have told you that there are 3 different types of carbon atom, one with no hydrogens attached (singlet at $\delta_C = 82$ ppm), one with one hydrogen (doublet at 74 ppm) and one with two hydrogens (triplet at 50 ppm). So we have C, CH, and $CH_2$.

The proton spectrum shows three different types of hydrogen, and as the broad resonance at 3.9 ppm disappears upon shaking with $D_2O$ you should have inferred this to be a labile proton, possibly on a hydroxyl group, OH. The intensities of the resonances are in the ratio of 1 : 1 : 2 from high to low field and, remembering the evidence for CH and $CH_2$ from the carbon-13 spectrum, we can take this ratio as representing the actual numbers of protons. Thus, summing up the proton information so far we have: OH, CH, and $CH_2$. We might also note that the $CH_2$ resonance at 4.4 ppm is coupled to the CH resonance at 2.5 ppm as these resonances are a doublet and triplet respectively.

Combining the carbon-13 and proton spectral data this structure must have: C, CH, $CH_2$ and OH. The mass of these fragments comes to 56 the known molecular mass. All that is required is to put these bits together. To do this a closer look at the precise chemical shifts involved for both the carbon-13 and proton spectra should have suggested that the compound has a carbon-carbon triple bond. In particular the carbon-13 shifts at 74 ppm and 82 ppm suggest $\equiv C-H$ and $R-C\equiv$ respectively. Thus the only possible structure for this material is:

$HO-CH_2-C\equiv C\text{-}H$   prop-2-ynol (propargyl alcohol)

Don't worry if you didn't get this one. This problem has fooled many an experienced interpreter. Probably the triple bond caught you out, and this is where more data from, say, ir spectra would have been useful.

***********************************

**SAQ 4.1a**

When making an *S/N* determination using the methylene quartet of a 1% v/v solution of ethylbenzene a peak-to-peak noise amplitude of 2.2 cm was found. The average signal amplitude was 17.5 cm. Which of the following is the *S/N* for the instrument?

(*i*) 7.95

(*ii*) 19.9

(*iii*) 38.5

(*iv*) 0.39

**Response**

Answer (*ii*) is correct

I hope you did not decide (*iv*) was correct. That answer should have been obviously wrong. If a signal could be measured at all then it must be greater than the noise and *S/N* must be greater than 1. After all, such a ratio implies that there is more noise than signal, and if that is the case then how do you know the signal amplitude—it is buried in the noise. Answer (*i*) comes from dividing the signal amplitude by the peak-to-peak noise. Now this is incorrect according to Eq. 4.1a and Eq. 4.1b. The *RMS* noise is required and that means dividing the peak-to-peak noise by 2.5. If this is done and Eq. 4.1a applied then an *S/N* of 19.9 is obtained, so answer (*ii*) is correct. If you chose answer (*iii*) you probably were looking optimistically at the figures and multiplied the signal amplitude by the noise.

**********************************

**SAQ 4.1b**

Which of the following are reasons for the low sensitivity of nmr spectroscopy? Circle T (true) or F (false)

(*i*) There is a great deal of instrument electronic noise.

T / F

(*ii*) Spin-energy levels are too close to one another.

T / F

(*iii*) The sample might be too hot.

T / F

**Response**

All these statements are correct.

Statement (*i*) is true, because unfortunately amplification of the very weak changes in magnetisation associated with nmr signals does introduce noise and so limits sensitivity. There are various types of electronic noise, from the general 'white noise' of random signals to 'environmental' noise like mains spiking (which can occur when nearby instruments, or other electrical equipment, is switched on or off) and mains hum from the 50 Hz AC mains frequency. Good instruments have ways of coping with these latter types of noise.

Statement (*ii*) is also correct; it refers to the small difference in the populations of the spin states that are undergoing resonance. According to the Boltzmann Equation, which we first met in Part 1 and which is shown below, the energy difference ($\Delta E$) helps to determine the relative populations of the energy states. The smaller the energy difference the smaller the population difference because $\Delta E/kT$ becomes more nearly zero and so $N_2/N_1$ more nearly 1.

$$N_2/N_1 = 1 - \Delta E/kT$$

where $N_2$, $N_1$ are the populations of the upper and low spin states of a two spin 1/2 system; $\Delta E$ is the energy difference between the spin states, $k$ is the Boltzmann constant, and $T$ is the absolute temperature.

Statement (*iii*), however silly it sounds at first, is, nevertheless true. The *S/N* of a sample decreases if the sample is warmed up. The explanation is seen again by reference to the Boltzmann equation. As T increases so $\Delta E/kT$ becomes smaller and again $N_2/N_1$ tends to equal 1. So the *S/N* diminishes because the population difference has gone down. Conversely, if the temperature of the sample was to be lowered the *S/N* would increase. This is not a practical solution for improving *S/N* because of other factors, such as the adverse effect of reduced solubility and increased sample viscosity.

***********************************

**SAQ 4.2a**

You have decided to try some signal enhancement on your nmr signals. List and arrange in the correct order four pieces of equipment you are going to need before the signals emerge on a chart recorder.

**Response**

Your list should be something like:

analogue–digital converter

computer

computer software (programs)

digital–analogue converter

You need them in that order because the signal has to be converted into digitised form for the computer to process using the appropriate software. Perhaps you did not consider computer programs as equipment, but they are just as much a tool as is an nmr spectrometer. Having done the data manipulation the digitised signals are reconverted via the DAC into a variable output voltage for the chart recorder. There is, of course, nothing to stop you displaying the signals in other ways such as a graphical output to your computer's visual display unit.

*********************************

**SAQ 4.2b**

Match the following statements with the terms, boxcar averaging, ensemble averaging, and weighted digital filtering.

(*i*) Multiscan, real-time spectral averaging.

(*ii*) Signal-to-noise enhancement by a factor of $\sqrt{n}$.

(*iii*) Post-scan data smoothing.

(*iv*) Rapid multipoint averaging of slow scan signals.

**Response**

You are probably thinking that the text was bad enough for terminology without all these statements, but jargon is a feature of scientific communication. Besides it sounds quite impressive to talk about 'multiscan, real-time spectral averaging' instead of ensemble averaging. Thus statement (*i*) provides a concise definition of CATing or ensemble averaging.

Statement (*ii*) applies to all the digital filtering methods although the meaning of *n* varies. In boxcar averaging *n* is the number of points read and averaged in one boxcar; in CATing *n* is the number of scans; while in weighted digital filtering *n* is the number of points in the smoothing function.

'Post-scan smoothing', statement (*iii*) can only apply to the weighted digital filtering methods because all the data needs to have been collected before this method can be used, hence 'post-scan'.

Statement (*iv*) gives a concise definition of the boxcar technique where a number of readings are rapidly measured and averaged. By the way, the term 'boxcar' is the American word for wagons in a freight train. The spectrum is thus seen as a train of boxcars—quite a good analogy.

************************************

**SAQ 4.2c**

You estimate that *S/N* enhancement by a factor of sixty-four is necessary to get useful nmr data from the spectrum of a particular sample. Which *one* of the following combinations of digital filtering techniques is going to be the most useful?

(*i*) A 16 point boxcar on each of a four-fold ensemble average followed by a 16 point Savitsky–Golay treatment.

(*ii*) A 32 point boxcar with a 4 scan CAT followed by a 32 point weighted filter.

(*iii*) A 16 point boxcar with 32 ensemble averages, then an 8 point Savitsky–Golay smooth.

(*iv*) A 128 point boxcar with a 4 scan CAT followed by an 8 point post scan smooth.

⟶

**Response**

If you worked out the arithmetic correctly you might have felt some frustration here because three of the options will give a factor of sixty-four enhancement. So we will have to use other criteria to decide between these three. First let us dispose of answer (*i*). Using the $\sqrt{n}$ factor mentioned in the text we can see that a 16 point boxcar would give a factor of 4 enhancement. A four fold ensemble average would give a 2 times enhancement, and the Savitsky–Golay treatment would give a four fold enhancement. So all told the sequence of digital filtering in (*i*) would give, theoretically $4 \times 2 \times 4 = 32$ times enhancement.

Using the same procedure answer (*ii*) gives: $\sqrt{32} \times \sqrt{4} \times \sqrt{32} = 64$; answer (*iii*): $\sqrt{16} \times \sqrt{32} \times \sqrt{8} = 64$; and answer (*iv*): $\sqrt{128} \times \sqrt{4} \times \sqrt{8} = 64$. So how can we choose between them? You might reject (*iii*) on the grounds that 32 scans are required for the CATing portion and that would take a long time. Answer (*ii*) might also be rejected because of the large number of points in the Savitsky–Golay function. This would probably lead to considerable line broadening and loss of resolution.

So answer (*iv*) is left. The options within it are quite reasonable, a 4 scan ensemble average would not take an excessive amount of time, and 8 point smooth should not affect resolution unduly, and a 128 point boxcar is within the range of modern ADC's and computers. Beware though, the theoretical enhancement is unlikely to be reached.

***********************************

**SAQ 4.3a**

From the list of words/phrases fill in the blanks in the following brief description of FT nmr.

FT nmr uses ... (1) ... of radiofrequency radiation to excite spin-active nuclei. The decay of magnetisation with time, the ... (2) ... is recorded and stored in a computer. The procedure is repeated as often as necessary and the resulting ... (3) ... ... (4) ... transformed to an ... (5) ... .

**Words/phrases:**

time-domain spectrum, frequency-domain spectrum, pulses, free induction decay, noise averaged.

**Response**

Unlike CW nmr FT nmr uses *pulses* of radiation. So blank (1) should have been 'pulses'. Blank (2) is 'free induction decay'. The magnetisation induced in the sample must decay with respect to time and this leads to the time-domain spectrum. Repeated pulsing and data collection will give an ensemble average so the noise is reduced. Thus (3) and (4) are 'noise averaged' and 'time-domain spectrum' respectively. Blank (5) must then be the 'frequency-domain spectrum', but remember it takes *time* to record the spectrum first.

***********************************

**SAQ 4.3b**

Are the following statements about free induction decay spectra correct or incorrect? Circle T (true) or F (false).

(*i*) FID's contain all the nmr information except coupling constants.

T / F

(*ii*) FID's cannot be interpreted because they are always too noisy.

T / F

(*iii*) Fourier transformation of an FID involves simple mathematics.

T / F

**Response**

Statement (*i*) is untrue. If there is spin–spin coupling in an nmr system then the time-domain spectrum, the FID, will contain that information. I know the text does not specifically mention coupling constants, but reference was being made to a simple one peak signal in Fig. 4.3c. Sometimes nmr spectra are extremely complex and both the time and frequency-domain spectra can, in a sense, contain too much information. Very recent developments in FT nmr allow chemical shift, relaxation time, or coupling constant information to be selectively extracted from the FID. This field is known as 2-D nmr spectroscopy, the 2-D meaning two dimensional. However, it is beyond the scope of this work.

Statement (*ii*) is false. FID spectra can certainly look noisy, but that is not really the reason for not being able to interpret them. The decay spectrum is made up from a pattern of interacting frequencies—an interference pattern. The complexity of the interference pattern is the reason why we, usually, cannot interpret FIDs. If you've not come across the idea before a useful analogy is the pattern on the surface of a pond when you toss a handful of pebbles into the water.

If you threw one stone you would observe a set of concentric waves radiating from the point where the stone entered. If you tossed two stones two sets of waves would appear and where they ran into one another there would be an interference pattern set up. Some waves would reinforce one another making a larger wave, some would cancel each other out, most would interact in an intermediate way. A whole handful of pebbles would cause a very complex pattern to emerge, and die away with time. Fourier transformation of that wave pattern would give you the exact places on the pond's surface where the stones had entered the water. Now, if you substitute in the analogy resonance frequencies for stones hitting the water you have a fair description of what an FID is.

Statement (*iii*) is false. To judge its truth we have to judge our mathematical abilities. No doubt some professional mathematicians would not be too distressed by Fourier transformation, but certainly at this level the mathematics is complex and beyond the scope of this work.

****************************************

**SAQ 4.4a**

> Make a list of at least six ideas, techniques, or things that have been referred to in Part 4 by their initials, eg *RMS* —Root Mean Square. (And you cannot count that one!)

**Response**

Your list should have the following:

*S/N* – signal-to-noise ratio;

CAT – computer of average transients;

FT – Fourier transformation;

FID – free induction decay;

ADC – analogue digital converter;

DAC – digital analogue converter;

and, of course, nmr—nuclear magnetic resonance.

***********************************

**SAQ 4.4b**

Which of the following statements about electronic noise are correct? Circle T (true) or F (false).

(*i*) Noise can never be completely eliminated from an nmr spectrum.

T / F

(*ii*) Useful information can be hidden in noise.

T /F

(*iii*) Noise deforms nmr signals.

T / F

(*iv*) Noise tends to average to zero if nmr signals are repetitively scanned and accumulated.

T / F

**Response**

All of these statements are true. Electronic noise can be a big problem in nmr spectroscopy and, unfortunately, it can never be completely eliminated from a spectrum. Of course, you may have sufficient sample in proton or fluorine-19 work so that you do not have to run your instrument at a level where the noise is apparent. But, as we have seen, even moderately dilute solutions at a 1% level have a lot of noise in their nmr spectra.

Useful information can certainly be hidden in noise. A single FID of a carbon-13 spectrum would appear to be all noise. Yet sufficient accumulations reveal the resonances that were hidden.

Noise does deform or distort nmr signals as a glance at any noisy spectrum will show. However, signal enhancement techniques can also cause deformations, particularly broadening of the peaks. It is part of the price that has to be paid to extract information. Noise is such a big problem that perhaps nmr spectroscopists should all be members of the Noise Abatement Society!

Repetitive scanning of the signals in both FT and CW nmr spectroscopy does tend to average noise to zero, but can never get rid of it completely. You continue the averaging techniques until you get an acceptably low noise level.

***********************************

**SAQ 5.1a**

An inexperienced nmr operator decided to analyse a commercial APC tablet for the relative quantities of aspirin and phenacetin. Not having a mortar and pestle to hand he crushed the tablet on a filter paper with a spatula and having placed the resulting lumps in a small flask he extracted them with carbon tetrachloride. Then he poured the carbon tetrachloride solution directly into an nmr tube. Nmr analysis gave him a ratio of 1 : 2 aspirin to phenacetin. List at least three errors that the operator made.

**Response**

As the APC tablet in question was a commercial sample we can assume that it had the correct 1 : 1 mol ratio of aspirin to phenacetin. So the 1 : 2 ratio observed was clearly incorrect. The operator's errors were:

(*i*) not crushing the tablet completely (a mortar and pestle are vital).

(*ii*) not using the correct solvent, $CCl_4$ being too non-polar and favouring the phenacetin; and

(*iii*) not filtering the extract. Any solid contaminants would degrade the quality of the nmr spectrum and so also any integrations.

The moral of this tale is that you must check everything if an unexpected result occurs.

***********************************

**SAQ 5.2a**

A sample of chateau-bottled claret (a red Bordeaux wine) analysed by nmr spectroscopy gave integration values of 876.0 for the hydroxyl resonance and 52.0 for the methylene resonance of the ethanol. Which of the following represents the strength of the wine?

(*i*) 6.57 : 1 water : ethanol mass ratio

(*ii*) 14.0% ethanol by volume

(*iii*) 16.8 : 1 water : ethanol mole ratio

(*iv*) 6.39 : 1 water : ethanol volume ratio

**Response**

A wine connoisseur might not have had too much trouble with this question even if he/she knew nothing about nmr spectroscopy.

Answer (*ii*) is correct.

Following the method outlined in the text you should have made a correction to the hydroxyl resonance integration for the contribution of the ethanol's hydroxyl group. As you know that two protons from ethanol give an integration of 52 then one proton would be 26. So the water to ethanol mole ratio is:

(876–26) : 52.0

ie, 850 : 52

16.3 : 1

So answer (*iii*) is wrong. In fact 16.8 is the ratio of total hydroxyl to ethanol, 876 : 52

To obtain the mass : mass ratio, you should have multiplied by the molecular masses, giving

15300 : 2392   water : ethanol

ie, 6.39 : 1

Answer (*i*) is therefore wrong as well. If you had this answer you might have used the wrong molecular masses, or relied on the incorrect mole ratio.

To obtain the volume to volume ratio of water to ethanol you need to divide the mass ratio by the appropriate densities. Thus

water : ethanol (volume : volume) = 6.39/1 : 1/0.96

ie, 6.39 : 1.04

6.14 : 1

So answer (*iv*) is incorrect, it is, as noted above the mass ratio not the volume ratio.

To obtain the % ethanol by volume you should have carried out the following:

$$\% \text{ ethanol} = \frac{\text{ethanol}}{\text{water} + \text{ethanol}} \times 100\%$$

$$= \frac{1}{7.14} \times 100 = 14.0\% \text{ v/v}$$

So answer (*ii*) is correct. You might note that this is a reasonable alcoholic content for such a wine. It might be termed 'full bodied'.

************************************

**SAQ 5.3a**

The amount of alcohol in a small sample of beer was determined by nmr spectroscopy using the method of standard additions. From the following results decide whether the brewer, who claimed that there was at least 4% alcohol present, was a charlatan or a paragon of business practice.

Each nmr sample was prepared by taking 0.50 $cm^3$ of the beer, adding the appropriate mass of ethanol and making up with $H_2O$ to 1.00 $cm^3$.

| | Sample | | | | |
|---|---|---|---|---|---|
| | 1 | 2 | 3 | 4 | 5 |
| Mass of Ethanol Added (mg) | 0.0 | 10.0 | 20.0 | 50.0 | 100.0 |
| Integration for $CH_3$ Resonance | 5 | 11 | 16 | 30 | 54 |

**Response**

You will have noticed that the brewer was being rather imprecise. He did not state whether the alcohol content was 4% by volume or by mass. (Well, he wouldn't would he!). Nevertheless, standard additions can help. A graph of mass of ethanol added against integration for these results is shown below. Extrapolation suggests that the original sample must have contained about 10 mg of ethanol. Now this came from 0.50 $cm^3$ of beer which was diluted by a factor of 2 so the original beer must have had about 40 mg in 1 $cm^3$. As we can take the mass of 1.00 $cm^3$ of beer to be about 1 g (after all it is mostly water) then the percentage ethanol by mass is about 4.0%. So on this basis the brewer is barely correct. Even taking account of densities the answer comes out to 4.2% so the claim is justified (just!).

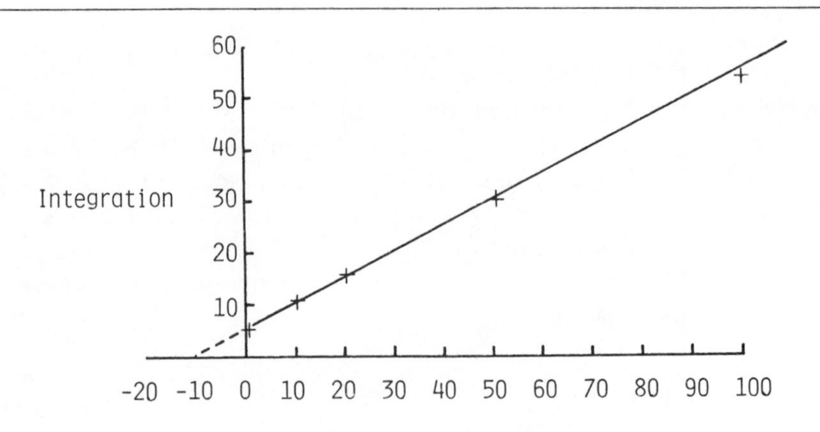


**Fig. 5.3d.** *Standard additions graph for ethanol in beer determination*

You might wonder why we do not use a direct integration of the ethanol resonance to the water resonance as we did for the whisky calculation. We could have, but measuring large integrations and comparing them to small ones in nmr spectroscopy introduces errors. The standard addition method gives a number of data points to use in calculation and increases the reliability of the answer.

***********************************

**SAQ 5.4a**

Oleic acid is a long chain carboxylic acid which contains one —CH=CH— group as part of the hydrocarbon chain. There are no elements present other than C, H, and O. In a sample submitted for nmr analysis these alkenic resonances had an integration of 14 units while the total integration for all the *rest* of the protons was found to be 222 units. Which of the following is the most likely molecular mass for oleic acid? ⟶

**SAQ 5.4a (cont.)**

(*i*) 256

(*ii*) 254

(*iii*) 282

(*iv*) 255

**Response**

Answer (*iii*) is correct.

The integration value for the alkenic resonances allows us to give a reasonably precise value for one proton, 7, ie half the alkenic value. This means that the total remaining integration of 222 is equivalent to 31.7 protons or 32 to the nearest integer. Now these 32 protons must be accounted for in the following way:

3 protons for the terminal $CH_3$ (mass 15)

1 proton for the carboxylic acid $CO_2H$ (mass 45)

28 protons for the chain of $CH_2$ groups
(14 $CH_2$ groups of mass 14 gives a total mass of 196)

So the total molecular mass is $15 + 45 + 196 + 26 = 282$
(the 26 comes from the $-CH{=}CH-$ group).

So according to the nmr results answer (*iii*) is correct. If you thought (*i*) was the one then you probably forgot to add in the molecular mass of the alkenic portion; similarly, answer (*ii*) is minus the carbonyl group ($-CO-$), while you should have realised answer (*iv*) is incorrect because of its odd molecular mass. Compounds having only C, H and O in them must have even molecular mass.

***********************************

**SAQ 5.4b**

Ethylene/vinyl acetate copolymers (EVAs) are used extensively in the plastic industry and have the general structure and composition shown below.

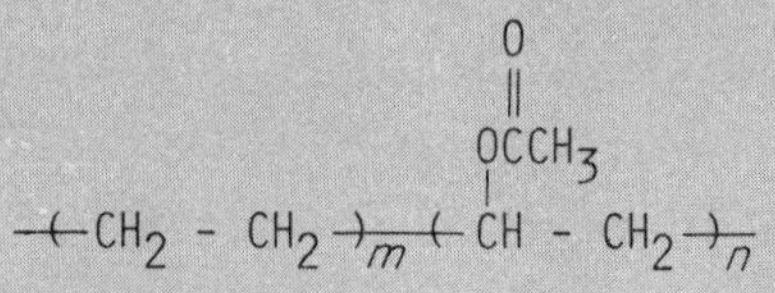

$$\left(CH_2 - CH_2\right)_m \left(\underset{\substack{|\\ OC(=O)CH_3}}{CH} - CH_2\right)_n$$

ethylene units    vinyl acetate units

The acetyl protons and the methylene protons give separate resonances whose areas can be measured by integration. (The methine proton also has a separate signal, but this does not affect the calculations). In one particular sample of EVA the integration ratio of methylene protons to acetyl protons was found to be 3 : 1. Which of the following is the percentage by weight of vinyl acetate in the sample?

(*i*) 84.3%

(*ii*) 15.7%

(*iii*) 75.4%

(*iv*) 25.0%

**Response**

Answer (*i*) is correct.

You might feel you need some of that claret to help you here, but a little reflection shows this to be a variation on the ethanol/water theme.

Now the methylene signal contains a contribution from the $CH_2$ protons of the vinyl acetate units as well as from the ethylene units. If there are '$m$' ethylenes and '$n$' vinyl acetates in the polymer chain then the methylene resonance must come from $4m + 2n$ protons. The acetyl resonance comes from $3n$ protons.

Thus the observed integration ratio of 3:1 is the same as $(4m + 2n):3n$ ie, $(4m + 2n)/3n = 3$.

$$\therefore \quad 4m + 2n = 9n$$

$$4m = 7n$$

$$\therefore \quad m = 7/4n$$

ie, there are 7/4 vinyl acetate units (molecular mass = 86) for every ethylene unit (molecular mass = 28).

So the percentage composition by weight for vinyl acetate is

$$\frac{\text{mass of vinyl acetate}}{\text{mass of vinyl acetate} + \text{mass of ethylene}}$$

$$= \frac{7/4 \times 86}{7/4 \times 86 + 1 \times 28} \times 100\% = 84.3\%$$

So answer (*i*) is correct. Answer (*ii*) is the percentage of ethylene by weight, answer (*iii*) is the value you would have obtained if you took '$m$' and '$n$' to be equal and answer (*iv*) is what you would have got by taking integration values as they stood without taking molecular masses into account.

OK, you were right. That was a difficult problem and you might have to check through it several times before you can understand it completely. Whether you were correct or not award yourself a bottle of claret for valiant effort.

**********************************

**SAQ 5.5a**

The ideal internal standard for quantitative nmr spectroscopy probably does not exist. For each of the following give at least one disadvantage for its use as an internal standard

(*i*) TMS

(*ii*) Iodoform

(*iii*) Sodium ethanoate ($Na^+CH_3CO_2^-$)

(*iv*) Malonic acid ($CH_2(CO_2H)_2$)

**Response**

You may have noticed that TMS, the usual nmr reference was missing from Fig. 5.5a. The reason why TMS is a poor reference is quantitative nmr is its volatility. This causes difficulties in making up accurate standard solutions and also in measurements on solutions that have been kept for any length of time. TMS evaporates from solutions fairly rapidly.

Iodoform is a useful standard, but its main disadvantage is that it is difficult to remove from samples. Precious samples would have to be chromatographed to get rid of iodoform. It is also light sensitive (especially in $CDCl_3$ solution).

The disadvantage of sodium ethanoate is that it is a salt and can only be used in water or other highly polar solvents. (Of course, another disadvantage of all the other compounds in this question is that they are insoluble in water and so are restricted to less polar media). Sodium ethanoate is slightly hygroscopic so care must be taken to use freshly dried material.

Malonic acid suffers from being an acid, like ethanoic acid, and so could react with many compounds. The most obvious type of reaction would be protonation which would alter chemical shifts as well as integration values.

**SAQ 5.5b**

The following results were obtained in a purity check on a sample of $CD_3OD$.

0.0601 g of $CHI_3$ dissolved in 0.8528 g $CD_3OD$ gave the following resonance intensities:

$CHI_3$, 10.3; $CD_3OH$, 8.64; $CD_2HOD$, 11.9.

Which of the choices listed below is the most likely proton content (mole %) of this sample?

(*i*) 1.31%

(*ii*) 0.55%

(*iii*) 0.76%

(*iv*) 1.25%

**Response**

Answer (*i*) is correct.

We can work out the proton content exactly as outlined in the text. First, we calculate the number of moles of standard and solvent;

0.0601 g $CHI_3$ ($M_r$ = 384) = $1.565 \times 10^{-4}$ mol

0.8528 g $CD_3OD$ ($M_r$ = 36) = $2.369 \times 10^{-2}$ mol

$\therefore$ If $1.565 \times 10^{-4}$ mol of standard gives an integration of 10.3

then there is $(8.64/10.3) \times 1.565 \times 10^{-4}$
$= 1.313 \times 10^{-4}$ mol of $CD_3OH$

and $(11.90/10.3) \times 1.565 \times 10^{-4}$
$= 1.808 \times 10^{-4}$ mol of $CD_2HOD$

$\therefore$ percentage $CD_3OH$
$= (1.313 \times 10^{-4}/2.369 \times 10^{-2}) \times 100 = 0.55\%$

and percentage $CD_2HOD$
$= (1.808 \times 10^{-4}/2.369 \times 10^{-2}) \times 100 = 0.76\%$

Total H content = 0.55 + 0.76 = 1.31 mol %.

So answer (*i*) is correct. You might have guessed that these are the results from a second determination on the $CD_3OD$ sample used in the text. All good analyses should be determined in duplicate. If you chose answers (*ii*) or (*iii*) then you were not counting the *total* proton content, only the $CD_3OH$ or $CD_3HOD$ amounts. Answer (*iv*) is the same as the value obtained in the text—a highly unlikely answer for a second determination.

***********************************

**SAQ 5.6a**

Fill in the gaps in the following description from the list below.

The determination of primary and secondary ... (1) ... of polyether polyols by fluorine-19 nmr spectroscopy depends on the ... (2) ... in the trifluoroacetyl esters of the polyether polyols sensing structural differences ... (3) ... distant. Chemical shift differences of ... (4) ... result between the primary and secondary esters, and these signals can be easily ... (5) ... so allowing quantitative determinations.

(*i*) four bonds

(*ii*) five bonds

(*iii*) 3–4 ppm →

**SAQ 5.6a (cont.)**

(*iv*) integrated

(*v*) hydroxyl content

(*vi*) 0.3–0.4 ppm

(*vii*) fluorine nuclei

**Response**

This paragraph is a short description of the determination of primary and secondary *hydroxyl content* in polyether polyols. So space (1) should have option (*v*). The method relies on *fluorine nuclei* being very sensitive to structural changes, thus space (2) should have option (*vii*). These nuclei can 'see' differences up to eight bonds away so sensing structural changes either four or five bonds distant is not a problem. With trifluoroacetyl esters the structural difference is not a problem. With trifluoroacetyl esters the structural difference is five bonds away from the fluorine nuclei so space (3) should have option (*ii*). The chemical shift differences that are induced are about *0.3–0.4 ppm* so space (4) would have option (*vi*). (Option (*iii*) is expecting too much even from fluorine). The resulting signals are easily *integrated* so the last space has option (*iv*).

***********************************

**SAQ 5.6b**

A sample of a polyether polyol treated with trifluoroacetic anhydride gave the following set of digital integration values for the primary and secondary trifluoroacetyl esters. What is the percentage of primary hydroxyl?

Average of 5 integrations for primary resonance = 12.4 ⟶

**SAQ 5.6b (cont.)**

Average of 5 integrations for secondary resonance = 108.2

(*i*) 10.3%

(*ii*) 11.5%

(*iii*) 89.7%

(*iv*) 23.2%

**Response**

This question should be straightforward.

$$\% \text{ primary} = \frac{\text{Integration for primary}}{\text{Total Integration}} \times 100\%$$

$$= \frac{12.4}{120.2} \times 100 = 10.3\%$$

So answer (*i*) is correct. Answer (*ii*) is the percentage derived from taking a direct ratio of the integrations. Answer (*iii*) is the percentage of secondary hydroxyl, while answer (*iv*) is completely spurious.

************************************

# Units of Measurement

For historic reasons a number of different units of measurement have evolved to express quantity of the same thing. In the 1960s, many international scientific bodies recommended the standardisation of names and symbols and the adoption universally of a coherent set of units—the SI units (Système Internationale d'Unités)—based on the definition of five basic units: metre (m); kilogram (kg); second (s); ampere (A); mole (mol); and candela (cd).

The earlier literature references and some of the older text books, naturally use the older units. Even now many practicing scientists have not adopted the SI unit as their working unit. It is therefore necessary to know of the older units and be able to interconvert with SI units.

In this series of texts SI units are used as standard practice. However in areas of activity where their use has not become general practice, eg biologically based laboratories, the earlier defined units are used. This is explained in the study guide to each unit.

Table 1 shows some symbols and abbreviations commonly used in analytical chemistry; Table 5 is a glossary of abbreviations used in this particular text. Table 2 shows some of the alternative methods for expressing the values of physical quantities and the relationship to the value in SI units.

More details and definition of other units may be found in the *Manual of Symbols and Terminology for Physicochemical Quantities and Units*, Whiffen, 1979, Pergamon Press.

**Table 1** *Symbols and Abbreviations Commonly used in Analytical Chemistry*

| | |
|---|---|
| Å | Angstrom |
| $A_r(X)$ | relative atomic mass of X |
| A | ampere |
| $E$ or $U$ | energy |
| $G$ | Gibbs free energy (function) |
| $H$ | enthalpy |
| J | joule |
| K | kelvin (273.15 + $t$ °C) |
| $K$ | equilibrium constant (with subscripts p, c, therm etc.) |
| $K_a, K_b$ | acid and base ionisation constants |
| $M_r(X)$ | relative molecular mass of X |
| N | newton (SI unit of force) |
| $P$ | total pressure |
| $s$ | standard deviation |
| $T$ | temperature/K |
| $V$ | volume |
| V | volt (J $A^{-1}$ $s^{-1}$) |
| $a$, $a$(A) | activity, activity of A |
| $c$ | concentration/ mol $dm^{-3}$ |
| e | electron |
| g | gramme |
| $i$ | current |
| s | second |
| $t$ | temperature / °C |
| | |
| bp | boiling point |
| fp | freezing point |
| mp | melting point |
| $\approx$ | approximately equal to |
| $<$ | less than |
| $>$ | greater than |
| e, exp($x$) | exponential of $x$ |
| ln $x$ | natural logarithm of $x$; $\ln x = 2.303 \log x$ |
| log $x$ | common logarithm of $x$ to base 10 |

**Table 2** *Alternative Methods of Expressing Various Physical Quantities*

1. **Mass (SI unit : kg)**

$$g = 10^{-3}\ kg$$
$$mg = 10^{-3}\ g = 10^{-6}\ kg$$
$$\mu g = 10^{-6}\ g = 10^{-9}\ kg$$

2. **Length (SI unit : m)**

$$cm = 10^{-2}\ m$$
$$Å = 10^{-10}\ m$$
$$nm = 10^{-9}\ m = 10Å$$
$$pm = 10^{-12}\ m = 10^{-2}\ Å$$

3. **Volume (SI unit : $m^3$)**

$$l = dm^3 = 10^{-3}\ m^3$$
$$ml = cm^3 = 10^{-6}\ m^3$$
$$\mu l = 10^{-3}\ cm^3$$

4. **Concentration (SI units : mol $m^{-3}$)**

$$M = mol\ l^{-1} = mol\ dm^{-3} = 10^3\ mol\ m^{-3}$$
$$mg\ l^{-1} = \mu g\ cm^{-3} = ppm = 10^{-3}\ g\ dm^{-3}$$
$$\mu g\ g^{-1} = ppm = 10^{-6}\ g\ g^{-1}$$
$$ng\ cm^{-3} = 10^{-6}\ g\ dm^{-3}$$
$$ng\ dm^{-3} = pg\ cm^{-3}$$
$$pg\ g^{-1} = ppb = 10^{-12}\ g\ g^{-1}$$
$$mg\% = 10^{-2}\ g\ dm^{-3}$$
$$\mu g\% = 10^{-5}\ g\ dm^{-3}$$

5. **Pressure (SI unit : N $m^{-2}$ = kg $m^{-1}$ $s^{-2}$)**

$$Pa = Nm^{-2}$$
$$atmos = 101\ 325\ N\ m^{-2}$$
$$bar = 10^5\ N\ m^{-2}$$
$$torr = mmHg = 133.322\ N\ m^{-2}$$

6. **Energy (SI unit : J = kg $m^2$ $s^{-2}$)**

$$cal = 4.184\ J$$
$$erg = 10^{-7}\ J$$
$$eV = 1.602 \times 10^{-19}\ J$$

**Table 3** *Prefixes for SI Units*

| Fraction | Prefix | Symbol |
|---|---|---|
| $10^{-1}$ | deci | d |
| $10^{-2}$ | centi | c |
| $10^{-3}$ | milli | m |
| $10^{-6}$ | micro | $\mu$ |
| $10^{-9}$ | nano | n |
| $10^{-12}$ | pico | p |
| $10^{-15}$ | femto | f |
| $10^{-18}$ | atto | a |

| Multiple | Prefix | Symbol |
|---|---|---|
| 10 | deka | da |
| $10^{2}$ | hecto | h |
| $10^{3}$ | kilo | k |
| $10^{6}$ | mega | M |
| $10^{9}$ | giga | G |
| $10^{12}$ | tera | T |
| $10^{15}$ | peta | P |
| $10^{18}$ | exa | E |

**Table 4** *Recommended Values of Physical Constants*

| Physical constant | Symbol | Value |
|---|---|---|
| acceleration due to gravity | $g$ | $9.81\ m\ s^{-2}$ |
| Avogadro constant | $N_A$ | $6.022\ 05 \times 10^{23}\ mol^{-1}$ |
| Boltzmann constant | $k$ | $1.380\ 66 \times 10^{-23}\ J\ K^{-1}$ |
| charge to mass ratio | $e/m$ | $1.758\ 796 \times 10^{11}\ C\ kg^{-1}$ |
| electronic charge | $e$ | $1.602\ 19 \times 10^{-19}\ C$ |
| Faraday constant | $F$ | $9.648\ 46 \times 10^{4}\ C\ mol^{-1}$ |
| gas constant | $R$ | $8.314\ J\ K^{-1}\ mol^{-1}$ |
| 'ice-point' temperature | $T_{ice}$ | 273.150 K exactly |
| molar volume of ideal gas (stp) | $V_m$ | $2.241\ 38 \times 10^{-2}\ m^3\ mol^{-1}$ |
| permittivity of a vacuum | $\epsilon_o$ | $8.854\ 188 \times 10^{-12}\ kg^{-1}\ m^{-3}\ s^4\ A^2\ (F\ m^{-1})$ |
| Planck constant | $h$ | $6.626\ 2 \times 10^{-34}\ J\ s$ |
| standard atmosphere pressure | $p$ | $101\ 325\ N\ m^{-2}$ exactly |
| atomic mass unit | $m_u$ | $1.660\ 566 \times 10^{-27}\ kg$ |
| speed of light in a vacuum | $c$ | $2.997\ 925 \times 10^{8}\ m\ s^{-1}$ |

**Table 5** *Glossary and Abbreviations used in nmr Spectroscopy*

| | |
|---|---|
| ADC | Analogue–digital converter |
| CAT | Computer of average transients |
| CW | Continuous wave |
| DAC | Digital–analogue converter |
| E | Energy |
| FID | Free induction decay |
| FT | Fourier transform |
| Hz | Hertz |
| MHz | Megahertz |
| NMR (nmr) | Nuclear magnetic resonance |
| ppm | Parts per million |
| Rf | Radiofrequency |
| *RMS* | Root mean square |
| *S/N* | Signal-to-noise ratio |

**TERMS**

| | |
|---|---|
| Chemical shift | $\delta$ ppm ($10^{-6}$) |
| Coupling constant | $J$ Hz ($s^{-1}$) |
| Frequency | $\nu$ MHz ($10^6$ $s^{-1}$) |
| Magnetic field | $B_O$ Tesla or gauss<br>(1 gauss = $10^{-4}$ Tesla) |
| Relaxation time | $T_1$, $T_2$ s |
| Spin quantun number | $I$ |